PRACTICE PROBLEMS WITH SOLUTI

to accompany

College PHYSICS

Fifth Edition

RAYMOND A. SERWAY
James Madison University

JERRY S. FAUGHN
Eastern Kentucky University

Saunders College Publishing
Harcourt Brace College Publishers

Fort Worth Philadelphia San Diego New York Orlando Austin
San Antonio Toronto Montreal London Sydney Tokyo

Printed in the United States of America

ISBN 0-03-22488-8

901 023 765432

PREFACE

This workbook contains over 300 problems to supplement Serway and Faughn's *College Physics,* Fifth edition. These problems provide you with the additional practice needed to master the concepts of physics.

You should carefully plan your problem-solving method. First, read the problem several times and be sure you understand what is being asked. Second, look for key words that will help you interpret the problems and make assumptions regarding information you will need to solve the problem. Third, write down the information given in the problem– you might want to construct a table listing the quantities given and the quantities needed. Fourth, decide on your problem solving method and proceed with your solution. If you need help, refer to the worked examples in the text book.

If you do not get the correct answer on your first try, do not despair. Try again! With some practice you can become a skilled problem solver. You will also find that as your problem solving skills improve, physics will become more fun. You will find enjoyment in tackling the difficult problems, and you will feel a sense of accomplishment when your hard work translates into good grades.

Good luck and enjoy your physics class!

CONTENTS

1 INTRODUCTION

PROBLEMS

*Indicates intermediate level problems.

1. Show that the equation $v^2 = v_0^2 + 2ax$ is dimensionally correct, where v and v_0 represent velocities, a is acceleration and x is a distance.

2. Which of the equations below are dimensionally correct? (a) $v = v_0 + ax$ (b) $y = (2\text{ m})\cos(kx)$, where $k = 2\text{ m}^{-1}$.

3. Carry out the following arithmetic operations: (a) 756 minus 37.2, (b) 3.2 divided by 1.4577.

4. How many meters are there in the hundred-yard dash?

*5. Assume an oil slick consists of a single layer of molecules and that each molecule occupies a cube 1.0 mm on a side. Determine the area of an oil slick formed by 1.0 m^3 of oil.

*6. Estimate the number of Ping-Pong balls that would fit (without being crushed) into a room 4 m long, 4 m wide, and 3 m high. Assume that the diameter of a Ping-Pong ball is 3.8 cm.

7. You close out your $1,000,000 checking account at a bank, and take your money all in one-dollar bills. Estimate the height of your stack of money.

8. Two points in a rectangular coordinate system have coordinates (2.0, –4.0) and (–3.0, 3.0), where the units are in meters. Determine the distance between these points.

CHAPTER 1 SOLUTIONS

1. $v^2 = v_0{}^2 + 2ax$ has units of $(L/T)^2 = (L/T)^2 + (L/T^2)(L)$. All terms have units of L^2T^{-2}.

2. (a) Both v and v_o have units of L/T, but the term ax has units of

$$(L/T^2)(L) = \frac{L^2}{T^2}.$$

Thus, this equation is not dimensionally correct.

(b) The argument of the cosine term has no dimensions, and the units of the number 2 are those of m. Thus, the units are L on each side, and the equation is dimensionally correct.

3. (a) The answer is 719 because the number 756 carries no information beyond the decimal.
(b) 3.2/1.4577 must be rounded to 2.2 because 3.2 has only two significant figures.

4. $100 \text{ yds} = (100 \text{ yds})\left(\frac{3 \text{ ft}}{1 \text{ yd}}\right)\left(\frac{1 \text{ m}}{3.28 \text{ ft}}\right) = 91.4 \text{ m}$

*5. Volume of oil = 1.0 m^3

(area)(thickness) = 1.0 m^3

(area)($1.0\ \mu$m) = 1.0 m^3

$$\text{area} = \frac{1.0 \text{ m}^3}{1.0\ \mu\text{m}} = 10^6 \text{ m}^2\left(\frac{1 \text{ mile}}{1609 \text{ m}}\right)^2 = 0.39 \text{ sq. miles}$$

*6. The volume of the room is $4 \times 4 \times 3 = 48 \text{ m}^3$, while the volume of one ball is

$$\frac{4\pi}{3}(0.038 \text{ m})^3 = 2.38 \times 10^{-4} \text{ m}^3$$

Therefore, one can fit about $\frac{48}{2.38 \times 10^{-4}} = 200{,}000$ ping-pong balls in the room. As an aside, the actual number is much smaller than this because there will be a lot of space in the room that cannot be covered by balls. In fact, even in the best arrangement, the so-called "best packing fraction" is $\frac{\pi\sqrt{2}}{6} = 0.74$ so that at least 26% of the space will be empty. Therefore, the above estimate reduces to

$$200{,}000 \times 0.74 = 148{,}000.$$

7. Assume 200 bills per inch, and the height is found as

height = (number of bills)/(number of bills per inch)

$= (1 \times 10^6)/(200/\text{in}) = 5{,}000$ in, or 417 ft.

(Note: If the bills are placed end-to-end, they will stretch for about 100 miles.)

8. The x distance between these points is 5.0 m, and the y distance between these points is 7.0 m. The distance between the points is found from the Pythagorean theorem as

$$\sqrt{(5.0\ \text{m})^2 + (7.0\ \text{m})^2} = 8.6\ \text{m}$$

2 MOTION IN ONE DIMENSION

PROBLEMS

*Indicates intermediate level problems.

1. A football player makes a touchdown run of 100 yards in a time of 15.0 seconds. What was his average velocity in m/s during his run?

2. The position-time graph for a bug crawling along the x axis is shown in Figure 2.1. Determine whether the velocity is positive, negative, or zero for the times (a) t_1, (b) t_2, (c) t_3, (d) t_4.

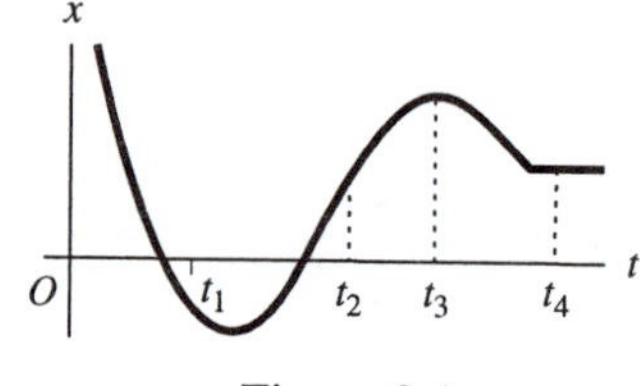

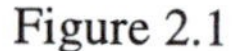

Figure 2.1

3. A jogger runs eastward in a straight line with an average speed of 2.0 m/s for 5.0 min and then continues with the average speed of 1.5 m/s for 2.0 min. (a) What is her total displacement? (b)What is her average velocity during this time?

4. It is found that the position of a model airplane as a function of time is given for a portion of time by $x = 2t^2$. Plot a graph of this equation between $t = 0$ and $t = 3.0$ s. From your graph find (a) the average velocity during this 3.0 s interval and (b) the instantaneous velocity at 2.0 s.

5. A shopper in a supermarket is in a great hurry. Plot a position-time graph for him as he moves along an aisle in a straight-line path. Use the following data and assume the origin of coordinates is at the initial position of the shopper. He moves from position 0 to –3.0 m at constant velocity in 1.0 s. He then moves from this position to +3.0 m at a constant velocity in 2.0 s. Finally, he pauses to catch his breath for 1.0 s. Use the graph you have plotted to find the average velocity during the total time interval, and the instantaneous velocity at 0.50 s, 2.0 s, and 3.5 s.

6. A car traveling in a straight-line path has a velocity of +10.0 m/s at some instant. After 3.00 s, its speed is +6.00 m/s. What is the average acceleration in this time interval?

7. A car starts from rest and accelerates at 0.300 m/s^2. What is the speed of the car after it has traveled 25.0 m?

*8. A car, initially traveling at 20.0 m/s, accelerates at a uniform rate of 4.00 m/s^2 for a distance of 50.0 m. How much time is required to cover this distance?

9. A bicyclist starts down a hill with an initial speed of 2.0 m/s. She moves down the hill with a constant acceleration, arriving at the bottom with a speed of 8.0 m/s. If the hill is 12 m long, what is the acceleration of the bicyclist on the hill?

10. The tallest volcano in the Solar System is the 24.0-km-tall Martian volcano, Olympus Mons. Assume an astronaut drops a ball off the rim of the crater and that the free-fall acceleration remains constant throughout the ball's 24.0-km fall at a value of 3.70 m/s^2. (We assume the crater is as deep as the volcano is tall, which is not the case in nature.) Find (a) the time for the ball to reach the crater floor and (b) the velocity with which it hits. (In light of your answer for the velocity, does it seem reasonable that air resistance, even in Mars' thin atmosphere, can really be neglected in this problem?)

11. A ball is thrown downward from the top of a tall cliff with an initial speed of 10.0 m/s. Determine the velocity and acceleration of the ball at $t = 2.00$ s.

12. When a nurse squeezes a syringe, the liquid squirts 3.43 cm into the air. With what speed does the liquid emerge from the syringe?

13. A foul ball is hit into the stands at a baseball game. The ball rises to a height of 40 m and is caught by a fan at a height of 30 m as it drops back toward the field. What is its velocity in the vertical direction just befo it is caught?

14. The velocity of an object is shown as a function of time if Figure 2.2. Find the average acceleration of the object in the intervals (a) 0 to 1 s, (b) 1 s to 3 s, and (c) 3 s to 4 s. (d) Find the instantaneous acceleration at 5.0 s, 2 s, and 3.5 s.

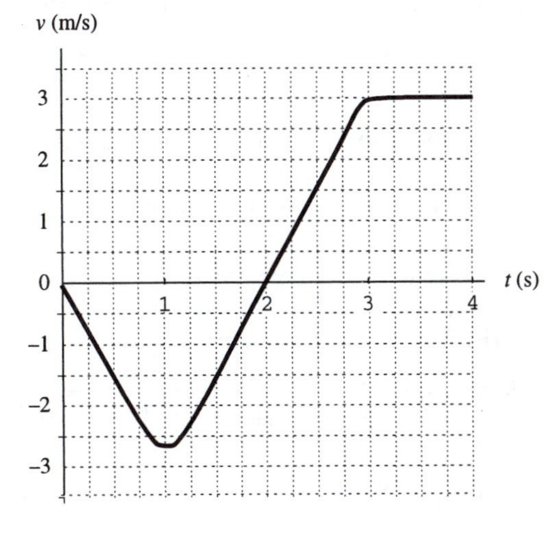

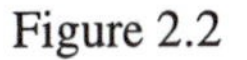
Figure 2.2

CHAPTER 2 SOLUTIONS

1. $\bar{v} = \Delta x/\Delta t = (100 \text{ yds}/15.0 \text{ s})(3 \text{ ft/ yd})(1 \text{ m}/3.28 \text{ ft}) = 6.10 \text{ m/s}$

2. Consider the slope of the line tangent to the graph at the point of interest.
   (a) $v < 0$ (slope is negative) (b) $v > 0$ (slope positive)
   (c) $v = 0$ (zero slope) (d) $v = 0$ (zero slope)

3. (a) The displacement in the first 5.0 minutes is

$$\Delta x_1 = \bar{v}_1(\Delta t)_1 = (2.0 \text{ m/s})(300 \text{ s}) = 600 \text{ m}$$

and the displacement in the next 2.0 min is

$$\Delta x_2 = \bar{v}_2(\Delta t)_2 = (1.5 \text{ m/s})(120 \text{ s}) = 180 \text{ m}$$

The total displacement $= \Delta x_1 + \Delta x_2 = 780$ m

(b) The average velocity

$$\bar{v} = (\Delta x)_{\text{total}}/(\Delta t)_{\text{total}} = 780 \text{ m}/420 \text{ s} = 1.9 \text{ m/s}$$

4. The plot of $x = 2t^2$ looks somewhat like that pictured below.

(a) The position of the plane at $t = 0$ is $x = 0$. At $t = 3.0$ s, the position is

$$x = 2(3.0)^2 = 18 \text{ m}$$

The average velocity is

$$\Delta x/\Delta t = 18 \text{ m}/3.0 \text{ s} = +6.0 \text{ m/s}$$

(b) The slope of the line tangent to the curve at $t = 2.0$ s gives the instantaneous velocity at that time. This slope is +8.0 m/s.

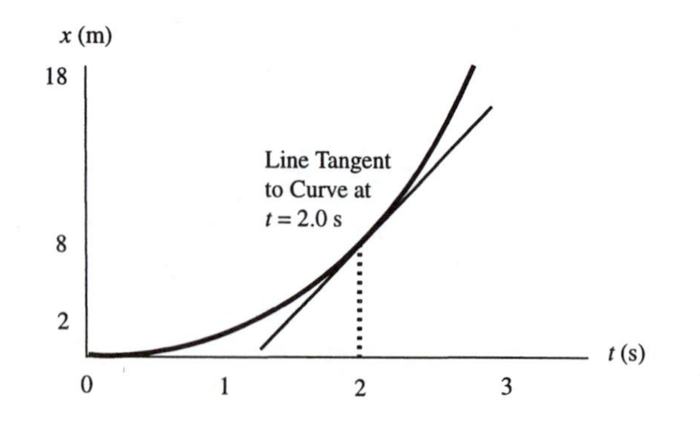

5. The position-time graph should look somewhat like that sketched below.

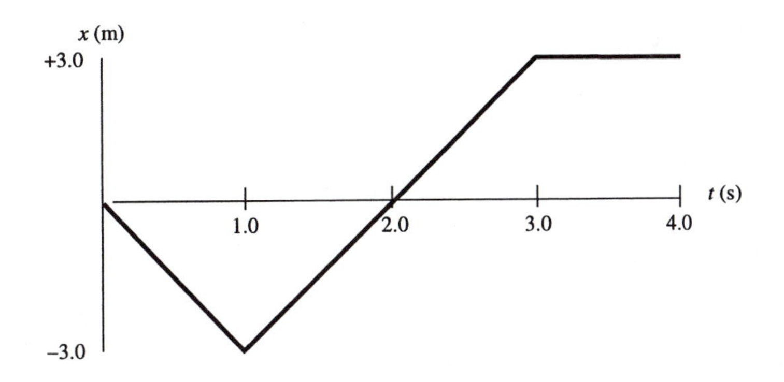

The average velocity during the total time interval is the total displacement divided by the total elapsed time.

$\bar{v}$ = + 3.0 m/ 4.0 s = 0.75 m/s

The instantaneous velocity at t = 0.50 s is the slope of the line between t = 0 and t = 1.0 s.

v = (−3.0 m − 0)/(1.0 s − 0) = −3.0 m/s

At t = 2.0 s, the instantaneous velocity is the slope of the line between t = 1.0 s and t = 3.0 s.

v = (+3.0 m − (−3.0 m))/2.0 s = +3.0 m/s

At t = 3.5 s, the instantaneous velocity is the slope of the line between t = 3.0 s and t = 4.0 s. This line has zero slope; thus, v = 0.

6. $\bar{a} = \Delta v/\Delta t = (v_f - v_i)/\Delta t$ = (6.00 m/s − 10.0 m/s)/3.00 s = (−4.00 m/s)/3.00 s = −1.33 m/s^2.

7. Use $v^2 = v_0^2 + 2ax$. We have

v^2 = 0 + 2(0.300 m/s^2)(25.0 m) and v = 3.87 m/s

*8. The final velocity of the car is found as

$v^2 = v_0^2 + 2ax$ = (20.0 m/s)^2 + 2(4.00 m/s^2)(50.0 m)

From which, v = 28.3 m/s

Then from the definition of acceleration, we find t as

$$t = \frac{v - v_0}{a} = \frac{28.3 \text{ m/s} - 20.0 \text{ m/s}}{4.00 \text{ m/s}^2} = 2.07 \text{ s}$$

9. Use $v^2 = v_0^2 + 2ax$.

$$a = \frac{v^2 - v_0^2}{2x} = \frac{(8.0 \text{ m/s})^2 - (2.0 \text{ m/s})^2}{2(12 \text{ m})} = +2.5 \text{ m/s}^2.$$

10. (a) The time to fall a distance of 24.0 km is found from $y = v_o t + \frac{1}{2}at^2$.

    Thus, -2.40×10^4 m $= 0 + \frac{1}{2}(-3.70 \text{ m/s}^2)t^2$. From which $t = 113.9$ s.

    (b) The velocity at this time is found from $v = v_o + at$.

    $v = 0 + (-3.70)(113.9 \text{ s}) = -421 \text{ m/s}$

    (Air resistance increases as the speed of an object increases. As a result, air resistance should become significant at these speeds.)

11. The acceleration of the ball is -9.80 m/s^2 even though it was thrown downward. The velocity of the ball is given by

    $v = v_0 + at = (-10.0 \text{ m/s}) - (9.80 \text{ m/s}^2)(2.00 \text{ s}) = -29.6 \text{ m/s}$

12. $v^2 = v_0{}^2 + 2gy$ becomes

    $0 = v_0{}^2 + 2(-9.80 \text{ m/s}^2)(3.43 \times 10^{-2} \text{ m})$

    From which, $v_0 = 0.820$ m/s

13. We shall use $v^2 = v_0{}^2 + 2ay$ in the vertical direction, and we shall select the origin of our coordinate system at the position of the maximum height reached by the ball. At this point, the initial velocity in the vertical direction is zero. The displacement from the maximum height, 40 m, to the point where it is caught, 30 m, is -10 m. Thus,

    $v^2 = 0 + 2(-9.80 \text{ m/s}^2)(-10 \text{ m})$

    and $v = -14$ m/s

14. We use $\bar{a} = \Delta v/\Delta t$

    (a) 0 to 1 s: $\bar{a} = \dfrac{((-3 \text{ m/s}) - 0)}{(1 \text{ s})} = -3 \text{ m/s}^2$

    (b) 1 s to 3 s: $\bar{a} = \dfrac{((3 \text{ m/s}) - (-3 \text{ m/s}))}{(2 \text{ s})} = 3 \text{ m/s}^2$

    (c) 3 s to 4 s: $\bar{a} = \dfrac{((3 \text{ m/s}) - 3 \text{ m/s})}{(1 \text{ s})} = 0$

    (d) The instantaneous acceleration is the slope of the tangent line drawn to the v vs t curve at the time of interest. At $t = 0.5$ s, the slope of this line is -3 m/s^2; at $t = 2$ s, the slope is 3 m/s^2 , and at $t = 3.5$ s, the slope is 0.

3 VECTORS AND TWO-DIMENSIONAL MOTION

PROBLEMS

*Indicates intermediate level problems.

1. While traveling along a straight interstate highway you notice that the mile marker reads 260. You travel until you reach the 150-mile marker and note that you are low on gas. You then retrace your path to the 175-mile marker. What is the magnitude of your resultant displacement from the 260-mile marker?

2. (a) What is the resultant displacement of a walk of 80 m followed by a walk of 125 m when both displacements are in the eastward direction? (b) What is the resultant displacement in a situation in which the 125-m walk is in the direction opposite the 80-m walk?

3. A submarine dives at an angle of 30° with the horizontal and follows a straight-line path for a total distance of 50 m. How far is the submarine below the surface of the water?

*4. A person going for a walk follows the path shown in Figure 3.1. The total trip consists of four straight-line paths. At the end of the walk, what is the person's resultant displacement measured from the starting point?

*5. A punter kicks a football at an angle of 30.0° with the horizontal at an initial speed of 20.0 m/s. Where should a punt returner position himself to catch the ball just before it strikes the ground?

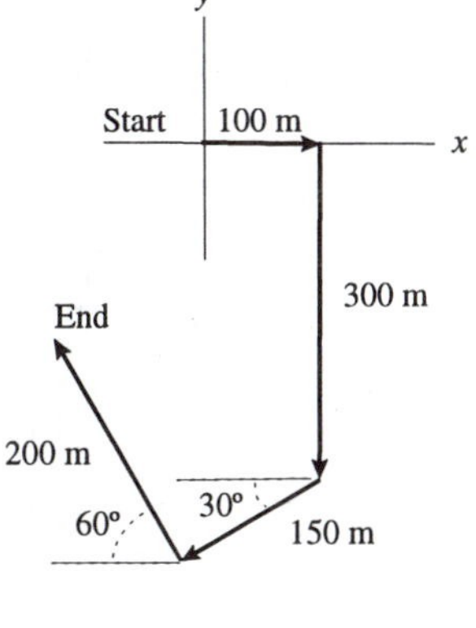

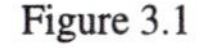
Figure 3.1

6. Two rowboats are heading toward one another. The speed of rowboat 1 is 0.511 km/h. The speed of rowboat 2 is 0.433 km/h. Assuming the direction of motion of rowboat 1 is positive, find the velocity of rowboat 2 as seen from rowboat 1.

7. A motorist traveling west on Interstate Route 80 at 80 km/h is being chased by a police car traveling at 95 km/h. (a) What is the velocity of the motorist relative to the police car? (b) What is the velocity of the police car relative to the motorist?

8. A river has a steady speed of 0.5 m/s. A student swims upstream a distance of 1.0 km and returns to the starting point. If the student can swim at a speed of 1.2 m/s in still water, how long does the trip take? Compare this with the time the trip would take if the water were still.

*9. A taxi driver travels due south for 10.0 km and then moves 6.00 km in a direction 30.0° north of east. Find the magnitude and the direction of the car's resultant displacement.

*10. While exploring a cave, a spelunker starts at the entrance and moves the following distances. She goes 75.0 m north, 250 m east, 125 m at an angle 30° north of east, and 150 m south. Find the resultant displacement from the cave entrance.

*11. Determine the height h in Figure 3.2.

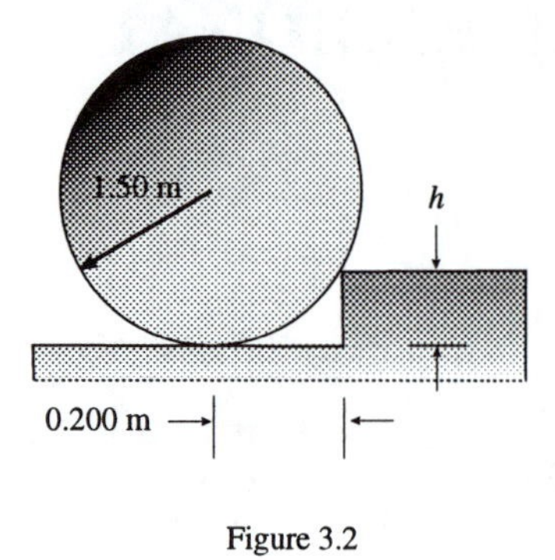

Figure 3.2

*12. Find the height h, the length of the side L, and the angle a of the triangle in Figure 3.3.

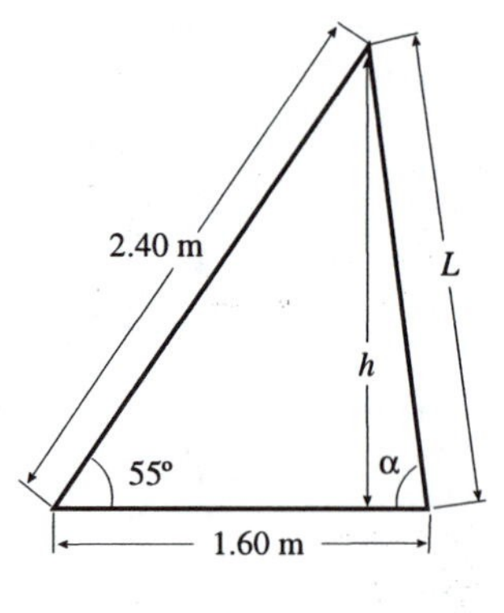

Figure 3.3

CHAPTER 3 SOLUTIONS

1. The displacement from the 260 mi marker to the 150 mi marker is 110 mi. The displacement vector from the 150 mi marker back to the 175 mi marker is 25 mi in the opposite direction to the first displacement. The vectors must be subtracted to give a resultant of 85 mi.

2. (a) When displacements are in the same direction, the vectors are added as algebraic quantities. Thus, an 80 m walk east followed by a 125 m walk also eastward gives a resultant displacement of 205 m eastward.
(b) The displacements are in opposite directions in this case, and thus, the vectors are subtracted to give a resultant of 45 m westward.

3. Diving at 30° below the horizontal for a distance of 50 m brings the sub to a distance h below the surface given by

$$h = (50\text{ m})(\sin 30°) = 25\text{ m}$$

*4. Let **A** be the vector corresponding to the 100 m displacement, **B** to the 300 m displacement, **C** to 150 m, and **D** to 200 m. The components are

$A_x = 100.0$ m $A_y = 0.0$
$B_x = 0.0$ $B_y = -300.0$ m
$C_x = -129.9$ m $C_y = -75.0$ m
$D_x = -100.0$ m $D_y = 173.2$ m

The resultant x component is –129.9 m, and the resultant y component is –201.8 m. From the Pythagorean theorem, we find

$$R = \sqrt{(R_x)^2 + (R_y)^2} = \sqrt{(-129.9\text{ m})^2 + (-201.8\text{ m})^2} = 240\text{ m}$$

and $\theta = \tan^{-1}\dfrac{R_y}{R_x} = \tan^{-1}\dfrac{-201.8}{-129.9} = \tan^{-1}(1.553)$

$$\theta = 237°$$

5. The time of flight of the football is found from $v = v_o + at$, applied in the vertical direction

$$t = \frac{-10.0\text{ m/s} - 10.0\text{ m/s}}{-9.80\text{ m/s}^2} = 2.04\text{ s}$$

The initial velocity in the x direction is found from

$$v_{ox} = v_o \cos\theta = (20.0\text{ m/s})\cos 30.0° = 17.3\text{ m/s}$$

The horizontal distance moved during this time can now be found as

$$x = v_{ox}t = (17.3\text{ m/s})(2.04\text{ s}) = 35.3\text{ m}$$

6. We have $\mathbf{v}_{1s} = 0.511$ km/h (velocity of boat 1 relative to the shore), and $\mathbf{v}_{2s} = -0.433$ km/h (velocity of boat 2 relative to the shore). Let $\mathbf{v}_{21}$ be the velocity of boat 2 relative to boat 1. Then

$$\mathbf{v}_{2s} = \mathbf{v}_{21} + \mathbf{v}_{1s}$$

or $\mathbf{v}_{21} = \mathbf{v}_{2s} - \mathbf{v}_{1s} = -0.433 \text{ km/h} - (0.511 \text{ km/h}) = -0.944 \text{ km/h}$.

7. (a) We have $\mathbf{v}_{Pe} = 95$ km/h (velocity of police relative to the earth), and $\mathbf{v}_{me} = 80$ km/h (velocity of motorist relative to the earth). We have

$$\mathbf{v}_{me} = \mathbf{v}_{mp} + \mathbf{v}_{pe}$$

or $\mathbf{v}_{mp} = \mathbf{v}_{me} - \mathbf{v}_{pe} = 80 \text{ km/h} - 95 \text{ km/h} = -15 \text{ km/h}$

(b) $\mathbf{v}_{pe} = \mathbf{v}_{pm} + \mathbf{v}_{me}$ gives

$$\mathbf{v}_{pm} = \mathbf{v}_{pe} - \mathbf{v}_{me} = 95 \text{ km/h} - 80 \text{ km/h} = 15 \text{ km/h}$$

8. $\mathbf{v}_{se}$ = the velocity of the swimmer relative to the earth.
$\mathbf{v}_{sW}$ = the velocity of the swimmer relative to the water.
$\mathbf{v}_{We}$ = the velocity of the water relative to the earth.

$$\mathbf{v}_{sW} = 1.2 \text{ m/s}, \text{ when going downstream}$$

$$\mathbf{v}_{se} = 1.2 \text{ m/s} + 0.50 \text{ m/s} = 1.7 \text{ m/s}$$

and the time $t_1 = \dfrac{1000\text{m}}{1.7 \text{ m/s}} = 588.2 \text{ s}$

$$\mathbf{v}_{sW} = -1.2 \text{ m/s}, \text{ when going upstream}$$

$$\mathbf{v}_{se} = -1.2 \text{ m/s} + 0.50 \text{ m/s} = -0.70 \text{ m/s}$$

and the time $t_2 = \dfrac{-1000 \text{ m}}{-0.70 \text{ m/s}} = 1428.6 \text{ s}$

Thus the time for the round trip = $t_1 + t_2 = 2017$ s (with flowing water). If the water were still, the time for the trip would be

$$t' = \frac{2(1000 \text{ m})}{1.2 \text{ m/s}} = 1667 \text{ s}$$

Thus, the percent increase in time due to the moving water is

$$\frac{2017 \text{ s} - 1667 \text{ s}}{1667 \text{ s}} 100\% = 21\% \text{ increase}$$

*9. The components are

	east(x component)		north (y component)
10 km:	0		−10.0 km
6 km:	5.20 km		+3.00 km
R_x =	+5.20 km	R_y =	−7.00 km

and from $\sqrt{(R_x)^2 + (R_y)^2}$, we find $R = 8.72$ km.

From $\tan\theta = \dfrac{R_x}{R_y} = -1.35$, we find $\theta = -53.4°$ or $\theta = 53.4°$ south of east.

*10. The components of the displacements are given below

	x comp		y comp
75 m	0.0		+75.0 m
250 m	+250.0		0.0
125 m	+108.0		+62.5
150 m	0.0		−150.0
R_x =	+358.0 m	R_y =	−12.5 m

and from the Pythagorean theorem

$$R = \sqrt{(358 \text{ m})^2 + (-12.5 \text{ m})^2} = 358 \text{ m}$$

and $\tan\theta = R_y/R_x = -0.004$ from which, $\theta = -2.00°$.

Thus, **R** = 358 m at 2° south of east.

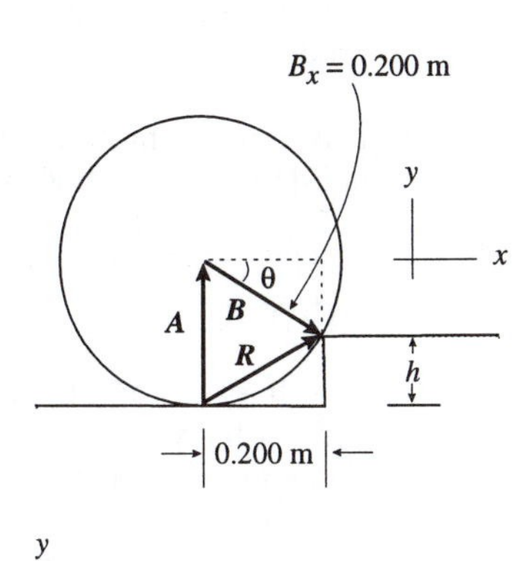

*11. First, find the angle θ: $B_x = B\cos\theta$, so $\cos\theta = \dfrac{0.200 \text{ m}}{1.50 \text{ m}} = 0.133$, and $\theta = 82.3°$. Then, realize that

$$h = R_y = A_y - B_y = 1.50 \text{ m} - (1.50 \text{ m})(\sin\theta)$$

$$= 1.50 \text{ m}(1 - \sin 82.3°) = 0.0134 \text{ m} = 1.34 \text{ cm}$$

*12. Recognize that $h = A_y = (2.40 \text{ m})\sin 55 = 1.97 \text{ m} = h = L_y$.

$L_x = 1.60 \text{ m} - A_x = 1.60 \text{ m} - (2.40 \text{ m})\cos 55 = 0.220 \text{ m}$

Then from the Pythagorean theorem,

$$L = \sqrt{(L_x)^2 + (L_y)^2} = \sqrt{(L_x)^2 + (h)^2} = 1.98 \text{ m}$$

and $\tan\alpha = \dfrac{h}{L_x} = \dfrac{1.97}{0.220} = 8.95$, and $\alpha = 83.6°$

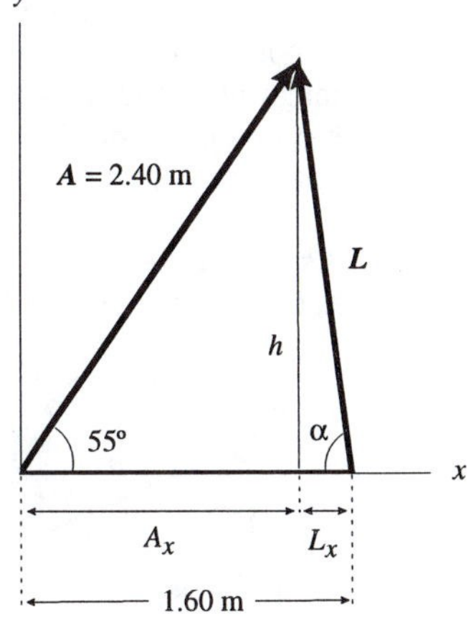

4 THE LAWS OF MOTION

PROBLEMS

*Indicates intermediate level problems.

1. An elevator accelerates upward at 1.50 m/s^2. If the elevator has a mass of 200 kg, find the tension in the supporting cable.

2. A student slides a 2000-N crate across a floor by pulling with a force of 750 N at an angle of 30.00° with the horizontal. If the crate moves at a constant speed, find the coefficient of sliding friction between the crate and the floor.

3. What is the weight in newtons of a 2.00-kg cannonball? What is its weight in pounds?

4. If a man weighs 900 N on Earth, what would he weigh on Jupiter, where the acceleration of gravity is 25.9 m/s^2?

5. A box weighing 20.0 N is to be pushed across a smooth floor with a horizontal force of 5.00 N. What acceleration will be produced?

6. It is amateur night at the tightrope walkers' convention, and a 600-N performer finds himself in the awkward position shown in Figure 4.1. If the angle between the rope and the horizontal is 8.00°, find the tension in the rope on either side of the performer.

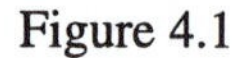
Figure 4.1

7. Two horses are pulling a sled across an icy road; assume that friction between the sled and the road is negligible. A farmer keeps the sled along the proper path by pulling on it with a rope attached at the point where the cables from the horses are attached. A noise startles the horse and they move towards opposite sides of the road as shown in Figure 4.2. What are the magnitude and the direction of the force that the farmer will have to exert on the sled to keep it moving along the road at at constant velocity?

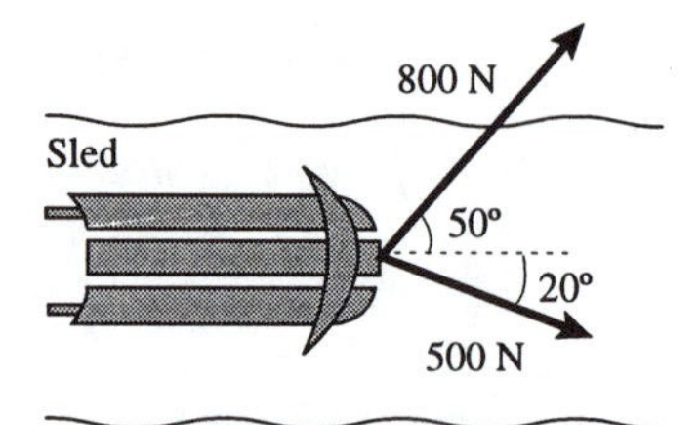

Figure 4.2

8. Assuming the hand grips the rope tightly in Figure 4.3, determine the force of the hand on the cord.

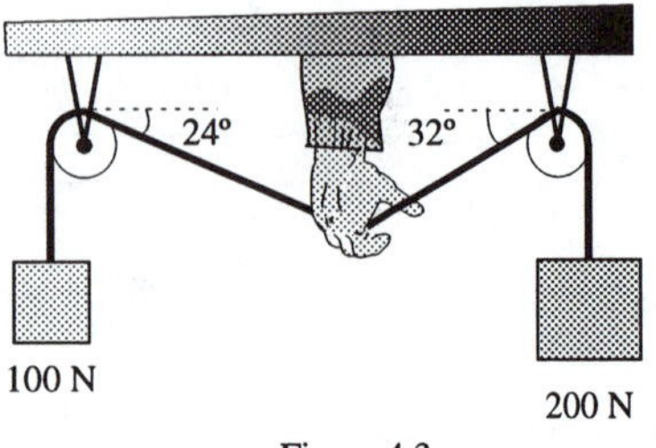

Figure 4.3

9. A crate is held at rest on a frictionless 53° ramp by a rope parallel to the incline. If the tension in the rope is 7000 N, what is the mass of the crate?

*10. A roller coaster starts its descent with an initial speed of 4.00 m/s. it moves through a distance of 135 ft along an incline that makes an angle of 40.0° with the horizontal. Neglect friction and find its speed at the bottom of the incline.

11. (a) A 10.0-kg box of books rests on a horizontal surface. If the coefficient of static friction box and floor is 0.400, find the maximum horizontal force that can be applied to the box before it slips. (b) The coefficient of kinetic friction is 20% less than that of static friction. If the force found in part (a) is exerted on the box once it is in motion, find the acceleration of the box.

12. A box weighing 5400 N is being pulled up an inclined plane that rises 1.30 m for every 7.50 m of length measured along the incline. If the coefficient of friction is 0.600, determine the force applied parallel to the incline necessary to move it up the incline at constant speed.

13. A furniture crate of mass 60.0 kg is at rest on a loading ramp that makes an angle of 25.0° with the horizontal. The coefficient of kinetic friction between the crate and the ramp is 0.200. What force, applied parallel to the incline, is required to push the crate up the incline at constant speed?

14. A tractor is pulling two logs in tandem. The logs are fastened together by a chain. Each log weighs 1750 N, and the logs are moving with constant velocity. The tractor exerts a force of 1600 N on the front log, and the same log is acted on by a force friction of 900 N. Find the tension in the chain between the logs and the force of friction on the back log.

15. A 3.00-kg ball is dropped from the roof of a building 176.4 m high. While the ball is falling to Earth, a horizontal wind exerts a constant force of 12.0 N on it. (a) How far from the building does the ball hit the ground? (b) How long does it take to hit the ground? (c) What is its speed when it hits the ground?

CHAPTER 4 SOLUTIONS

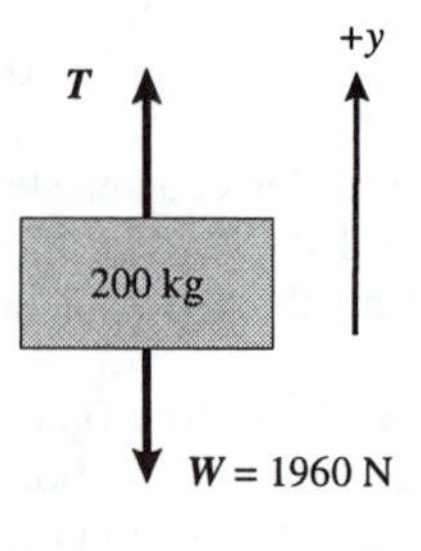

1. The forces on the elevator are the tension in the cable and its weight, 1960 N. The tension is found from the second law.

   $T - 1960\text{ N} = (200\text{ kg})(1.50\text{ m/s}^2)$

   $T = 2260\text{ N}$

2. (a) From $\Sigma F_x = 0$, we have $-f + (750\text{ N})\cos 30.0° = 0$

   gives $f = 650\text{ N}$

   From $\Sigma F_y = 0$, we have

   $N - 2000\text{ N} + (750\text{ N})\sin 30.0° = 0$

   From which, $N = 1625\text{ N}$

   Thus, $\mu_k = \dfrac{f}{N} = \dfrac{650}{1625} = 0.400$

3. $w = mg = (2.00\text{ kg})(9.80\text{ m/s}^2) = 19.6\text{ N}$

   $w = (19.6\text{ N})(0.225\text{ lb/ 1N}) = 4.41\text{ lb}$

4. Using $m = \dfrac{w}{g}$, we have $\dfrac{w_{\text{Jupiter}}}{g_{\text{Jupiter}}} = \dfrac{w_{\text{Earth}}}{g_{\text{Earth}}}$ (The mass is the same on both planets.)

   So, $w_{\text{Jupiter}} = w_{\text{Earth}}\left(\dfrac{g_{\text{Jupiter}}}{g_{\text{Earth}}}\right) = 900\text{ N}\left(\dfrac{25.9\text{ m/s}^2}{9.80\text{ m/s}^2}\right) = 2380\text{ N}$

5. The mass of the object is $m = w/g = (20.0\text{ N})/(9.80\text{ m/s}^2) = 2.04\text{ kg}$. Thus, the acceleration is given by $a = F_R/m = (5.00\text{ N})/(2.04\text{ kg}) = 2.45\text{ m/s}^2$.

6. $\Sigma F_x = 0$ becomes

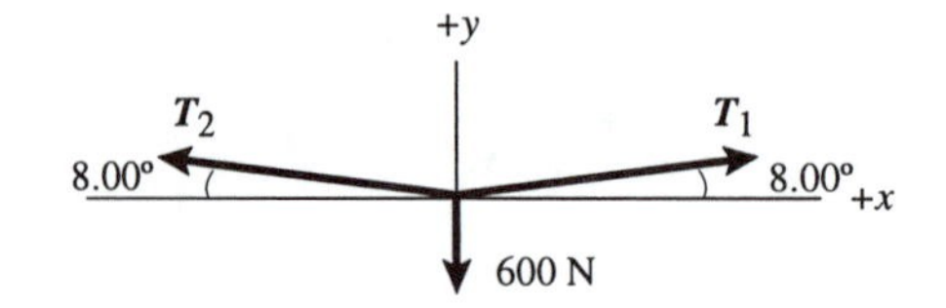

   $T_1 \cos 8.00° - T_2 \cos 8.00° = 0$ or $T_1 = T_2$

   $\Sigma F_y = 0$ becomes

   $T_1 \sin 8.00° + T_2 \sin 8.00° - 600\text{ N} = 0$

   or $2T_2 \sin 8.00° = 600\text{ N}$

   $T_2 = T_1 = 2160\text{ N}$

7. The resultant force in the x direction on the sled is given by

$$(800\ \text{N})\cos 50° + (500\ \text{N})\cos 20° = 984\ \text{N}$$

and the resultant force in the y direction is

$$(800\ \text{N})\sin 50° - (500\ \text{N})\sin 20° = 442\ \text{N}$$

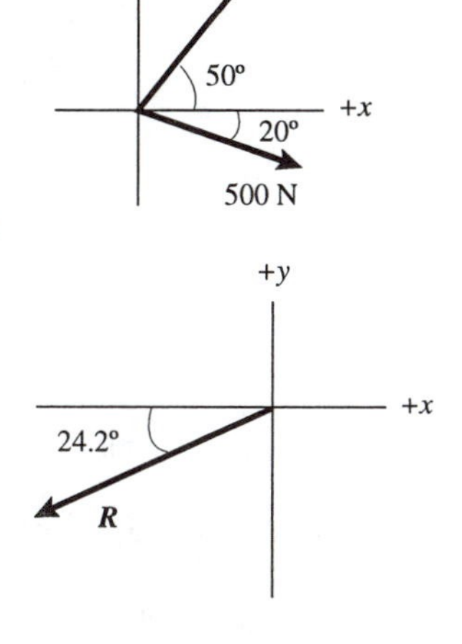

In order for the sled to move at constant velocity, the farmer will have to exert a force that will cancel these resultant component forces. Thus, his force will have components

$$R_x = -984\ \text{N} \text{ and } R_y = -442\ \text{N}$$

The magnitude of the farmer's force is found from the Pythagorean theorem as

$$\sqrt{(984\ \text{N})^2 + (442\ \text{N})^2} = 1080\ \text{N}$$

The angle at which this force must be exerted is $\tan\theta = \dfrac{442}{984} = 0.449$ and $\theta = 24.2°$

The angle is counterclockwise from the $-x$ direction. The force he must exert is

$$1080\ \text{N at } 24.2° + 180° = 204°$$

8. From $\Sigma F_x = 0$, we have

$$T_1 \cos 32.0° + F_x - T_2 \cos 24.0° = 0$$

or $F_x = -200\ \text{N} \cos 32.0° + 100\ \text{N} \cos 24.0° = -78.3\ \text{N}$

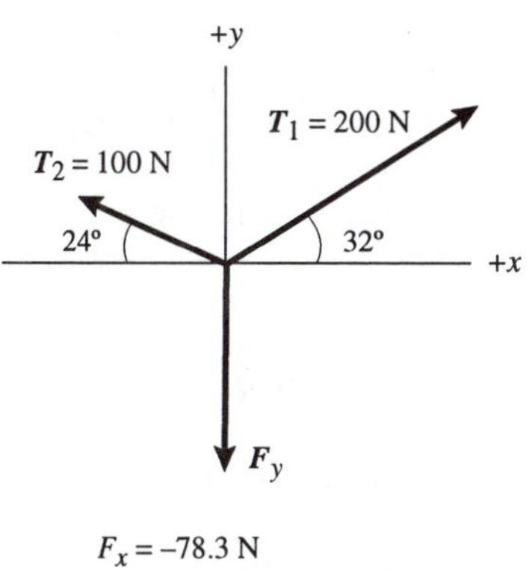

From $\Sigma F_y = 0$,

$$T_1 \sin 32.0° + T_2 \sin 24.0° - F_y = 0$$

or $F_y = 200\ \text{N} \sin 32.0° + 100\ \text{N} \sin 24.0° = 146.7\ \text{N}$

The force of the hand on the cord is found from the Pythagorean theorem as

$$\sqrt{(-78.3\ \text{N})^2 + (146.7\ \text{N})^2} = 166\ \text{N}$$

and $\tan\theta = \dfrac{F_x}{F_y} = 0.534 \qquad \theta = 28.1°$

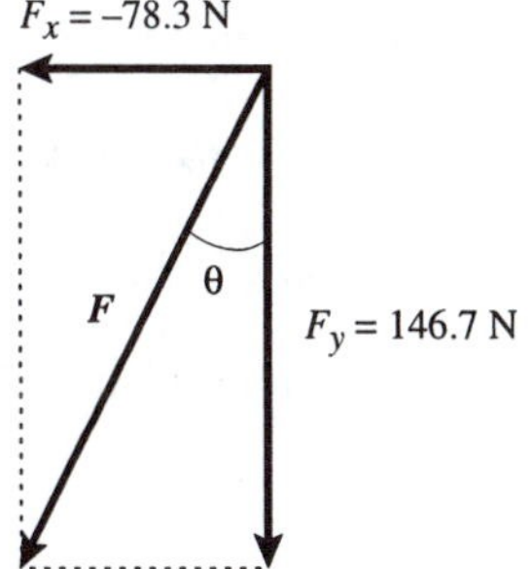

Thus the resultant force is 166 N, downward and to the left of vertical at an angle of 28.1°.

9. $\Sigma F_x = 0$ gives $T - mg \sin 53.0° = 0$ or, $7000\ \text{N} - 0.799\ mg = 0$

From which we find $m = 894$ kg

10. Let us choose the positive direction of the x axis down the incline. The acceleration along the 135 ft (41.1 m) incline is then found from the second law.

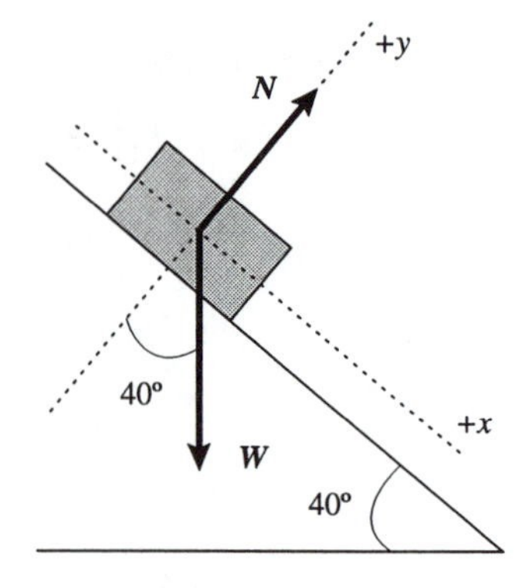

$$a_x = \frac{F_x}{m} = \frac{mg \sin \theta}{m} = g \sin \theta = (9.80 \text{ m/s}^2)(\sin 40.0°) = 6.30 \text{ m/s}^2$$

and

$$v_x{}^2 = v_{0x}{}^2 + 2a_x x = (4.00 \text{ m/s})^2 + 2(6.30 \text{ m/s}^2)(41.1 \text{ m})$$

giving

$$v_x = 23.1 \text{ m/s}$$

11. (a) When motion is impending, we know that

$$f_{\text{max}} = \mu_s N = 0.400(98.0 \text{ N}) = 39.2 \text{ N}$$

Thus, the maximum value of the force that can be exerted without causing movement is $F = 39.2$ N

(b) When slipping occurs, $f = \mu_k N$

but $\mu_k = 0.800\ \mu_s = 0.800(0.400) = 0.320$

Thus,

$$f = 0.320(98.0 \text{ N}) = 31.4 \text{ N}$$

Thus, if the applied force is maintained at 39.2 N as found in (a), we have

$$\Sigma F_x = ma_x$$

$$39.2 \text{ N} - 31.4 \text{ N} = (10.0 \text{ kg})a$$

and $a = 0.780 \text{ m/s}^2$.

12. The normal force is

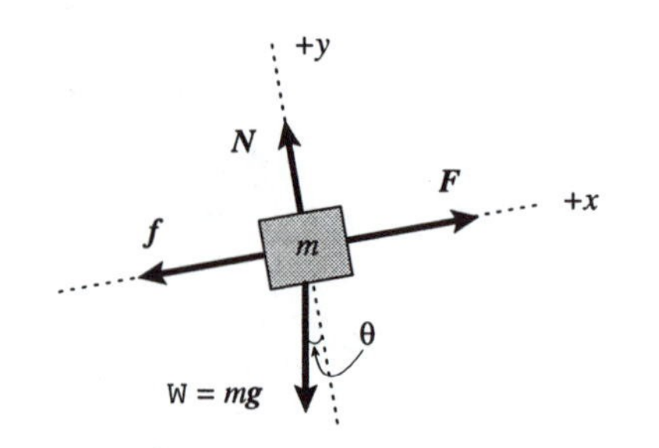

$$N = mg \cos \theta \quad \text{and} \quad f = \mu_k N$$

The angle θ is found as

$$\sin \theta = 1.30/7.50 = 0.173$$

From which, $\theta = 10°$

Let $+x$ be up the incline, and use $\Sigma F_x = 0$

$$F - f - w\sin \theta = 0$$

12. (continued)
Where F is the force required to pull the box up the incline.

With $\mu_k = 0.600$, $\theta = 10.0°$, and $w = 5400$ N, we find

$F = 4130$ N

13. See the solution to problem 12. The approach is identical, and we find

$F = f + w\sin\theta$

with $m = 60.0$ kg, $\mu_k = 0.200$, and $\theta = 25.0°$, we have

$F = 355$ N

14. Consider the free body diagram of log 1. It moves with a constant velocity, so

$T_1 - T_2 - f_1 = 0$, or

$T_2 = 1600\text{ N} - 900\text{ N} = 700\text{ N}$ (This is the tension in the chain between logs.

Now, consider the free body diagram of log 2:

$T_2 - f_2 = 0$, so, $f_2 = 700$ N (This is the force of friction on the back log.)

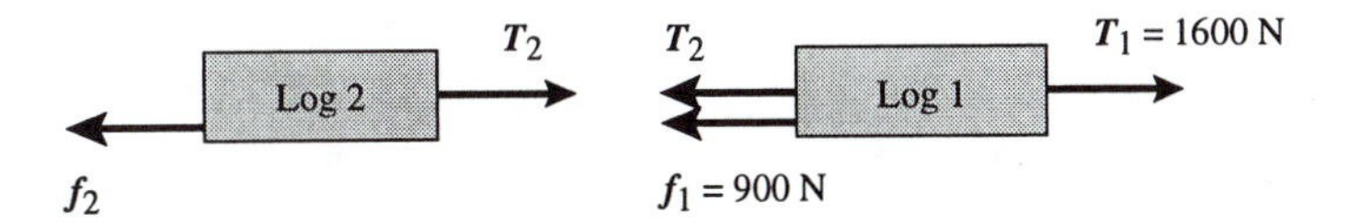

15. [(a) and (b)] Choose $y = 0$ at roof level with upward as positive. We have:

$a_y = -g = -9.80\text{ m/s}^2$, $v_{oy} = 0$, and

$$a_x = \frac{\text{wind force}}{m} = \frac{12.0}{3.00} = 4.00\text{ m/s}^2$$

When the ball reaches the ground ($y = -176.4$ m), we have

$-176.4\text{ m} = -4.90t^2$ or $t = 6.00$ s

The horizontal position is given by $x = v_{ox}t + \frac{1}{2}a_x t^2$.

Thus, at $t = 6.00$ s, $x = 0\ \frac{1}{2}(4.00\text{ m/s}^2)(6.00\text{ s})^2 = 72.0$ m.

(c) Using $v_y = -9.80t$ and $v_x = 4.00t$ gives

$v_y = -9.80(6.00) = -58.8$ m/s and $v_x = 4.00(6.00) = 24.0$ m/s

Then $v = \sqrt{v^2{}_x + v^2{}_y} = \sqrt{(-58.8)^2 + (24.0)^2} = 63.5$ m/s

5 WORK AND ENERGY

PROBLEMS

*Indicates intermediate level problems.

1. What is the kinetic energy of a 3000-kg car moving at 55.0 mi/h? How much heat, in joules, will be lost through friction in the brake linings when the car is brought to a stop?

2. A very good major league pitcher can throw a baseball at 100 mi/h, while an average pitcher can throw the ball at a speed of about 80.0 mi/h. If a baseball has a mass of 0.150 kg, find the kinetic energy of the ball in each case.

3. A gas molecule in the room you are sitting in may have a speed of about 50.0 m/s. If its mass is equal to 5.00×10^{-26} kg, what is its kinetic energy?

4. A 1200-kg vehicle is moving along a horizontal surface with a speed of 20.0 m/s. What work must be done by the brakes to bring the vehicle to rest in 20.0 s?

5. What is the gravitational potential energy relative to the ground of a 0.150-kg baseball at the top of a 100-m tall building?

6. A baseball player decides to show his skill by catching a 0.150-kg baseball dropped from the top of a 100-m tall building. What is the speed of the baseball just before it strikes his glove?

7. A 70.0-kg high-jumper leaves the ground with a speed of 6.00 m/s. How high can he leap?

8. A baseball player throws a baseball straight up into the air with an initial speed of 20.0 m/s. Find (a) the maximum height to which the ball rises, and (b) the speed of the ball when it is halfway up to its maximum height.

9. A woman weighing 500 N glides across some ice, starting her glide with a speed of 4.00 m/s. If the coefficient of friction between the skates and the ice is 0.115, how far does she go before coming to rest? Solve this problem by using the method of work-kinetic energy.

10. What average power is developed by an 800-N professor while running at constant speed up a flight of stairs rising 6.0 m if he takes 8.0 s to complete the climb?

11. A motor is used to pull an 80.0-kg skier along a horizontal surface at a constant speed of 2.00 m/s to enable him to learn to keep his balance. If the coefficient of kinetic friction between the skis and the surface is 0.150, what horsepower motor is required?

12. A machine lifts a 300-kg crate at constant speed through a height of 5.00 m in 8.00 s. Calculate the power output of the machine.

13. You have just found the "Lost Dutchman" gold mine, 80.0 ft below ground level. However, the mine is flooded, and you discover that in order to keep it dry enough to mine, you must pump water out at the rate of 60.0 lb every second. What is the minimum horsepower motor that can be used to perform this task?

14. A speedboat requires 130 hp to move at a constant speed of 15.0 m/s (≈33 mi/h). Calculate the resistive force due to the water at that speed.

15. Find the height from which you would have to drop this textbook so that it would have a speed of 10 m/s just before it hits the pavement.

16. A driver of a 7.50×10^3 N car passes a sign stating "Bridge Out 30 Meters Ahead." She slams on the brakes, coming to a stop in 10.0 s. How much work must be done by the brakes on the car if it is to stop just in time? Neglect the weight of the driver, and assume that the negative acceleration of the car caused by the braking is constant.

17. A 5.00-g bullet moving at 600 m/s penetrates a tree trunk to a depth of 4.00 cm. (a) Use work and energy considerations to find the average frictional force that stops the bullet. (b) Assuming that the frictional force is constant, determine how much time elapses between the moment the bullet enters the tree and the moment it stops moving.

CHAPTER 5 SOLUTIONS

1. The initial speed of the car is 55.0 mi/h = 24.6 m/s, and the initial kinetic energy is

$$KE_i = \frac{1}{2}mv^2 = \frac{1}{2}(3000 \text{ kg})(24.6 \text{ m/s})^2 = 9.07 \times 10^5 \text{ J}$$

The initial kinetic energy of the car must be dissipated in the brake linings to bring the car to a stop.

energy lost = 9.07×10^5 J

2. The speed of the ball thrown by the very good pitcher is 100 mi/h = 44.7 m/s. Similarly, the average pitcher's ball has a speed of 80.0 mi/h = 35.8 m/s. The kinetic energy of the fast pitch is

$$KE = \frac{1}{2}mv^2 = \frac{1}{2}(0.150 \text{ kg})(44.7 \text{ m/s})^2 = 150 \text{ J}$$

Using the same approach as above, the kinetic energy of the slower pitch is

$$KE = \frac{1}{2}mv^2 = \frac{1}{2}(0.150 \text{ kg})(35.8 \text{ m/s})^2 = 96.0 \text{ J}$$

3. The kinetic energy of the gas molecule is found as

$$KE = \frac{1}{2}mv^2 = \frac{1}{2}(5.00 \times 10^{-26} \text{ kg})(500 \text{ m/s})^2 = 6.25 \times 10^{-21} \text{ J}$$

4. The net work is equal to the change in kinetic energy, $W_{net} = \Delta KE$. Thus,

$$W_{net} = KE_f - KE_i = 0 - KE_i = -\frac{1}{2}mv_i^2 = -\frac{1}{2}(1200 \text{ kg})(20.0 \text{ m/s})^2 = -2.40 \times 10^5 \text{ J}.$$

5. $PE = mgy = (0.150 \text{ kg})(9.80 \text{ m/s}^2)(100 \text{ m}) = 147 \text{ J}$

6. We use conservation of mechanical energy with the zero level selected at the base of the building. We also know that the initial speed of the ball is zero.

$$\frac{1}{2}mv_i^2 + mgy_i = \frac{1}{2}mv_f^2 + mgy_f$$

$$0 + mg(100 \text{ m}) = \frac{1}{2}mv_f^2 + 0$$

The mass cancels, and we solve for v_f to find

$v_f = 44.3$ m/s

7. We use conservation of mechanical energy with the zero level for gravitational potential energy at ground level. The velocity at the maximum height, h, goes to zero, and we have

$$\frac{1}{2}mv_i^2 + mgy_i = \frac{1}{2}mv_f^2 + mgy_f$$

$$\frac{1}{2}m(6.00 \text{ m/s})^2 + 0 = 0 + mgh$$

Solve for h to find $h = 1.84$ m

8. (a) We choose the zero level for potential energy at ground level and note that at the maximum height, h, the velocity of the ball goes to zero. Thus, conservation of mechanical energy is written as

$$\frac{1}{2}mv_i^2 + mgy_i = \frac{1}{2}mv_f^2 + mgy_f$$

$$\frac{1}{2}m(20.0 \text{ m/s})^2 + 0 = 0 + mgh$$

Which gives $h = 20.4$ m

(b) At half the maximum height, 10.2 m, we find the speed as

$$\frac{1}{2}mv_i^2 + mgy_i = \frac{1}{2}mv_f^2 + mgy_f$$

$$\frac{1}{2}m(20.0 \text{ m/s})^2 + 0 = \frac{1}{2}mv^2 + mg(10.2 \text{ m})$$

giving, $v = 14.1$ m/s

9. $W_{nc} = \Delta KE + \Delta PE$

$$fs \cos 180° = 0 - \frac{1}{2}mv_i^2 + 0 - 0$$

$$(\mu_k mg)s(-1) = -\frac{1}{2}mv_i^2$$

$$(0.115)(9.80 \text{ m/s}^2)s = \frac{1}{2}(4.00 \text{ m/s})^2$$

$s = 7.10$ m

10. $W_{nc} = \Delta PE = mgh = (800 \text{ N})(6.0 \text{ m}) = 4.80 \times 10^3$ J

$\overline{P} = W/\Delta t = 4.8 \times 10^3 \text{ J}/8.0 \text{ s} = 600 \text{ W} = 0.80$ hp

11. If the skier is to be pulled at constant speed, the force applied by the motor is equal to the force of friction. Thus,

$$\overline{P} = fv = \mu_k mgv = 0.150(80.0 \text{ kg})(g)(2.00 \text{ m/s}) = 235 \text{ W} = 0.315 \text{ hp}$$

12. $P = \Delta W/\Delta t = mgh/\Delta t = (300 \text{ kg})g(5.00 \text{ m})/8.00 \text{ s} = 1.84 \times 10^3 \text{ W} = 2.46 \text{ hp}$

13. To lift 60.0 lbs of water through a height of 80.0 ft (at constant speed) requires a minimum amount of work given by,

$$W = Fs = (60.0 \text{ lbs})(80.0 \text{ ft}) = 4.80 \times 10^3 \text{ ft lbs}$$

To do this much work each second requires a power input of

$$P = \Delta W/\Delta t = 4.80 \times 10^3 \text{ ft lbs/s} = 8.73 \text{ hp}$$

14. $130 \text{ hp} = 9.70 \times 10^4 \text{ W}$

$P = Fv$, where F is the propulsive force required to overcome water resistance

$$F = \frac{P}{v} = \frac{9.70 \times 10^4 \text{ W}}{15.0 \text{ m/s}} = 6.47 \times 10^3 \text{ N (approximately equal to 1500 lb)}$$

15. Use conservation of mechanical energy, with the zero point of potential energy chosen to be at the top of the professor's head.

$$\frac{1}{2}mv_i^2 + mgy_i = \frac{1}{2}mv_f^2 + mgy_f$$

$$0 + \text{mgh} = \frac{1}{2}m(10 \text{ m/s})^2$$

The mass cancels, and we find

$$h = 5.1 \text{ m}$$

16. To determine the constant acceleration, we first use

$$v_f = v_o + at$$

with $v_f = 0$ to get $v_o = -at$.

Then $x = v_o t + \frac{1}{2}at^2 = -at^2 + \frac{1}{2}at^2 = -\frac{1}{2}at^2$, or

$$a = \frac{-2x}{t^2} = \frac{-2(30.0)}{(10.0)^2} = -0.600 \text{ m/s}^2$$

16. (continued)
The total force acting on the car is then

$$F = ma = \frac{7500}{9.80}(-0.600) = 459.2 \text{ N}$$

and the total work done by the brakes is

$$W = Fs\cos\theta = (459.2)(30.0)\cos 180° = -1.38 \times 10^5 \text{ J}$$

17. (a) The work due to friction is given by the work-energy theorem as

$$W_f = KE_f - KE_i = 0 - \frac{1}{2}mv_i^2 = -\frac{1}{2}(0.00500)(600)^2 = -900 \text{ J}$$

But this is also given by $W_f = -fD = -f(0.0400)$, so that

$$-0.0400f = -900$$

or $f = 2.30 \times 10^5$ N

(b) Using $v_f = v_i + at = v_i - \left(\frac{f}{m}\right)t = 0$, so

$$t = \frac{mv_i}{f} = \frac{(0.005)(600)}{22500} = 1.33 \times 10^{-4} \text{ s} = 0.133 \text{ ms}$$

6 MOMENTUM AND COLLISIONS

PROBLEMS

1. A 0.15-kg baseball is thrown with a speed of 10 m/s. (a) How much more momentum does it have if its speed is doubled? (b) How much more kinetic energy will it have?

2. The momentum of a 1500-kg car is equal to the momentum of a 5000-kg truck traveling with a speed of 15.0 m/s. What is the speed of the car?

3. A 0.100-kg ball is thrown with a speed of 20.0 m/s at an angle of 30.0° with the horizontal. Find the momentum of the ball (a) at its maximum height, and (b) just before it strikes the ground.

4. An 18,000-kg van is moving with a speed of 15 m/s. If Superman is to stop the van in 0.50 s, what average force must he exert on the van?

5. A 40.0-kg child standing on a frozen pond throws a 0.500 kg stone to the east with a speed of 5.00 m/s. Neglecting friction between child and ice, find the recoil velocity of the child.

6. A popcorn kernel initially at rest explodes into two pieces, one piece having twice the mass of the other. What is the ratio of the velocities of the fragments?

7. A 65.0-kg boy and 40.0-kg girl, both wearing skates, face each other at rest on a skating rink. The boy pushes the girl, sending her eastward with a speed of 4.00 m/s. Describe the subsequent motion of the boy. (Neglect friction.)

8. An object is moving so that its kinetic energy is 150 J and the magnitude of its momentum is 30.0 kg·m/s. Determine the mass and speed of the object.

9. An estimated force-time curve for a baseball struck by a bat is shown in Figure 6.1. From this curve, determine (a) the impulse delivered to the ball and (b) the average force exerted on the ball.

10. A bumper car at an amusement park ride traveling at 0.8 m/s collides with an identical car at rest. This second car moves away with a speed of 0.5 m/s. What is the velocity of the first car after the collision?

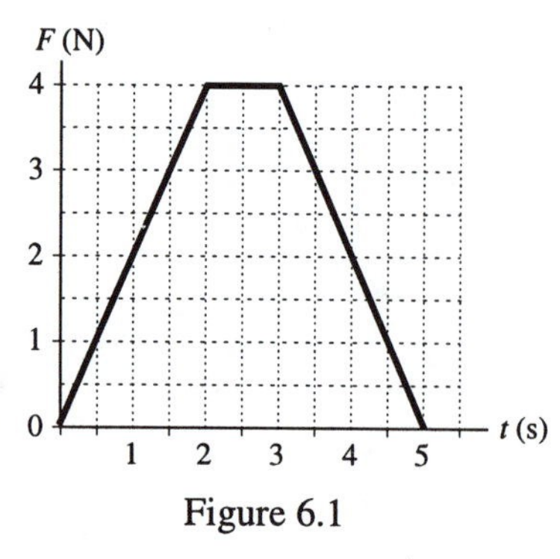

Figure 6.1

11. A 50.0-kg baseball pitching machine is placed on a frozen pond. If fires a 0.150-kg baseball at an angle of 30.0° wih the vertical at a speed of 20.0 m/s. What is the recoil velociy of the machine?

12. A 60.0-kg astronaut is on a space walk away from the shuttle when her tether line breaks! She is able to throw her 10.0-kg oxygen tank away from the shuttle with a velocity of 12.0 m/s to propel herself back to the shuttle. Assuming that she starts from rest (relative to the shuttle), determine the maximum distance she can be from the craft when the line breaks and still return within 60.0 s (the amount of time she can hold her breath).

13. A 1.20-kg skateboard is coasting along the pavement at a speed of 5.00 m/s when a 0.800-kg cat drops from a tree vertically downward onto the skateboard. What is the speed of the skateboard-cat combination?

14. A 0.11-kg tin can is resting on top of a 1.7-m-high fence post. A 0.0020-kg bullet is fired horizontally at the can. It strikes the can with a speed of 900 m/s, passes through it, and emerges with a speed of 720 m/s. When the can hits the ground, how far is it from the fence post? Disregard friction while the can is in contact with the post.

CHAPTER 6 SOLUTIONS

1. The original momentum and kinetic energy are $p_o = mv_o$, and $KE_o = \frac{1}{2}mv_o{}^2$. If the speed becomes $v = 2v_o$, the new momentum and kinetic energy are
(a) $p = mv = m(2v_o) = 2(mv_o) = 2p_o$ (momentum is doubled)

(b) $KE = \frac{1}{2}mv^2 = \frac{1}{2}m(2v_o)^2 = 4KE_o$ (kinetic energy is quadrupled)

2. We are given $p_c = p_t$

Therefore, $m_c v_c = m_t v_t$

and v_c = (5000 kg)(15.0 m/s)/1500 kg = 50.0 m/s

3. We have $v_{oy} = v_o \sin 30.0° = 10.0$ m/s

and $v_{ox} = v_o \cos 30.0° = 17.3$ m/s

(a) At the top of the arc, the velocity is solely in the x direction at 17.3 m/s. Thus,

$p = mv$ = (0.100 kg)(17.3 m/s) = 1.73 kgm/s (in the horizontal direction)

(b) At ground level the velocity of the ball has the same magnitude as that with which it was projected (20.0 m/s). Thus,

$p = mv$ = (0.100 kg)(20.0 m/s) = 2.00 kgm/s (at an angle of 30.0° below the horizontal.)

4. Use, $F\Delta t = \Delta p = mv_f - mv_i$.

$F(0.50 \text{ s}) = 0 - (1.8 \times 10^4 \text{ kg})(15 \text{ m/s})$

$F = -5.4 \times 10^5$ N

5. The initial momentum = 0. Therefore, the final momentum, p_f, must also be zero. We have, (taking eastward as the positive direction),

$p_f = (40.0 \text{ kg})(-v_c) + (0.500 \text{ kg})(5.00 \text{ m/s}) = 0$

$v_c = -6.25 \times 10^{-2}$ m/s (The child recoils westward.)

6. The momentum before the event is zero, and after, it is

$2m(-v_2) + m(v_1) = 0$ (where v_2 is the mass of the larger piece)

This reduces to $v_2 = \frac{v_1}{2}$

Or, the particle of twice the mass has a velocity which is 1/2 that of the less massive part. Also, the velocities have opposite directions.

7. $p_{after} = p_{before}$

   Thus, $(65.0\ \text{kg})v_{boy} + (40.0\ \text{kg})(4.00\ \text{m/s}) = 0$

   $v_{boy} = -2.46\ \text{m/s}$ (so the boy moves westward)

8. The kinetic energy is given by $KE = \frac{1}{2}mv^2$, and the magnitude of the momentum is $p = mv$. Dividing one equation by the other yields

   $$\frac{KE}{p} = \frac{v}{2}$$

   or $v = \frac{2KE}{p} = \frac{2(150)}{30.0} = 10.0\ \text{m/s}$

   Then, $m = \frac{p}{v} = \frac{30.0}{10.0} = 3.00\ \text{kg}.$

9. (a) Impulse = area under curve = 2 triangular areas of altitude 18,000 N and base of 0.75 s,

   or $\overline{F}\,\Delta t = 2\left(\frac{1}{2}(0.75\ \text{s})(18{,}000\ \text{N})\right) = 1.35 \times 10^4\ \text{N s}$

   (b) $\overline{F} = \frac{\text{impulse}}{\text{time}} = \frac{1.35 \times 10^4\ \text{N s}}{1.5\ \text{s}} = 9000\ \text{N}$ (approximately 2000 lb).

10. $p_{before} = M(.8\ \text{m/s}) + 0$; $p_{after} = Mv + M(0.5\ \text{m/s})$

    Use $p_{after} = p_{before}$ and solve for v, the velocity of the first car, gives

    $v = 0.3\ \text{m/s}$

11. Momentum is conserved only along the horizontal direction. We have

    $p_{after} = p_{before}$

    $-(50.0\ \text{kg})V + (0.150\ \text{kg})(20.0\ \text{m/s})(\cos 30.0°) = 0$

    $V = 5.20 \times 10^{-2}\ \text{m/s}$

    (V is the speed of the machine horizontally across the ice.)

12. The initial total momentum of the astronaut complete with oxygen tank is zero. After the astronaut throws her oxygen tank away at a speed of 12.0 m/s, we must have total momentum conserved, so that

    $(10.0)(-12.0) + 60.0\ v = 0$

    yielding $v = 2.00\ \text{m/s}$ as the speed of the astronaut toward the shuttle. Traveling at a constant speed for at most 60.0 s gives a maximum distance of $(60.0)(2.00) = 120\ \text{m}$.

13. $(p_x)_{after} = (p_x)_{before}$

    $(1.20 \text{ kg} + 0.800 \text{ kg})v = (1.20 \text{ kg})(5.00 \text{ m/s})$

    yielding $v = 3.00$ m/s

14. Let us use conservation of momentum to find the velocity V of the can immediately after the bullet passes through.

    $$(0.11 \text{ kg})(V) + (0.0020 \text{ kg})(720 \text{ m/s}) = (0.0020 \text{ kg})(900 \text{ m/s})$$

    or $V = 3.27$ m/s

    Now, treat the projectile motion of the can.

    We have: $y = v_{oy}t + \frac{1}{2}gt^2$

    or $-1.7 \text{ m} = 0 + \frac{1}{2}(-9.80 \text{ m/s}^2)t^2$

    which gives $t = 0.589$ s as the time to reach the ground.

    From its horizontal motion,

    $$x = v_{ox}t = (3.27 \text{ m/s})(0.589 \text{ s}) = 1.9 \text{ m}.$$

7 CIRCULAR MOTION AND THE LAW OF GRAVITY

PROBLEMS

*Indicates intermediate level problems.

1. Convert the following angles in radians to degrees: $\pi/3$, 1.2π, 3π.

2. Find the angular speed of the Earth about its axis in rad/s.

3. A particle moves in a circle 1.50 m in radius. Through what angle in radians does it rotate if it moves through an arc length of 2.50 m? What is this angle in degrees?

4. A grinding wheel, initially at rest, is rotated with constant angular acceleration $\alpha = 5.0$ rad/s^2 for 8.0 s. The wheel is then brought to rest, with uniform negative acceleration, in 10 rev. Determine the negative angular acceleration required and the time needed to bring the wheel to rest.

5. A centrifuge in a medical laboratory is rotating at an angular speed of 3600 rev/min. When switched off, it rotates 50.0 times before coming to rest. Find the constant angular deceleration of the centrifuge.

6. An airliner arrives at the terminal, and the engines are shut off. The rotor of one of the engines has an initial clockwise angular velocity of 2000 rad/s. The engine's rotation slows with an angular acceleration of magnitude 80.0 rad/s^2. (a)Determine the angular velocity after 10.0 s. (b) How long does it take the rotor to come to rest?

7. A race car travels in a circular track of radius 200 m. If the car moves with a constant speed of 80 m/s, find (a) its angular velocity and (b) its tangential acceleration.

8. The race car of Problem 7 increases its speed at a constant linear acceleration from 80 m/s to 95 m/s in 10 s. (a) Find the constant angular acceleration and (b) the angle the car moves through in this time.

9. A tire 0.500 m in radius rotates at a constant rate of 200 revolutions per minute. Find the speed and acceleration of a small stone lodged in the tread of the tire (on its outer edge).

10. The driver of a car traveling at 30.0 m/s applies the brakes and undergoes a constant negative acceleration of 2.00 m/s^2. How many revolutions does each tire make before the car comes to a stop, assuming that the car does not skid and that the tires have radii of 0.300 m?

11. Two students sitting in adjacent seats in a lecture room have weights of 600 N and 700 N. Assume that Newton's law of gravitation can be applied to these students and find the gravitational force that one student exerts on the other when they are separated by 0.500 m.

12. Communication satellites are placed at an altitude of about 25,000 mi above the surface of the Earth. Find the acceleration due to gravity at this altitude.

*13. A satellite of Mars has a period of 459 min. The mass of Mars is 6.42×10^{23} kg. From this information determine the radius of the satellite's orbit.

14. Young David experimented with slings before tackling Goliath. He found that he could develop an angular speed of 8.0 rev/s in a sling 0.60 m long. If he increased the length to 0.90 m, he could revolve the sling only six times per second. (a) Which angular speed gives the greater linear speed? (b) What is the centripetal acceleration at 8.0 rev/s? (c) What is the centripetal acceleration at 6.0 rev/s?

15. The downward motion of an elevator is controlled by a cable that unwinds from a cylinder of radius 0.20 m. What is he angular velocity of the cylinder when the downward speed of the elevator is 1.2 m/s?

16. A 60.0-cm diameter wheel rotates with a constant angular acceleration of 4.00 rad/s^2. It starts from rest at $t = 0$, and a chalk line drawn to a point, P, on the rim of the wheel makes an angle of 57.3° with the horizontal at this time. At $t = 2.00$ s, find (a) the angular speed of the wheel, (b) the linear velocity and tangential acceleration of P, and (c) the position of P.

CHAPTER 7 SOLUTIONS

1. The conversion relationship is π rad = 180°. To convert from rad to degrees, we use the conversion factor as follows.

$$\frac{\pi}{3}\text{ rad} = \frac{\pi}{3}\left(\frac{180°}{\pi\text{ rad}}\right) = 60°$$

Similarly, we find 1.2π rad = 216°, and 3π rad = 540°.

2. The earth turns through 2π rad in one day (86,400 s). Thus,

$$\omega = \frac{2\pi\text{ rad}}{86400\text{ s}} = 7.27 \times 10^{-5}\text{ rad/s}$$

3. $\theta = \frac{s}{r} = \frac{2.50\text{ m}}{1.50\text{ m}} = 1.67\text{ rad} = 95.7°$

4. First find the speed attained before the wheel begins to slowing down. With $\omega_o = 0$, we have

$$\omega = \omega_o + \alpha t = 0 + (5.0\text{ rad/s}^2)(8.0\text{ s}) = 40\text{ rad/s}$$

Next, consider the wheel as it comes to rest. Now $\omega_o = 40$ rad/s, and

$$\omega^2 = \omega_o^{\,2} + 2\alpha\theta \quad \text{becomes} \quad 0 = (40\text{ rad/s})^2 + 2\alpha(20\pi\text{ rad})$$

from which $\alpha = -13\text{ rad/s}^2$.

To obtain the time, use $\theta = \bar{\omega}t$ as

$$t = \frac{\theta}{\bar{\omega}} = \frac{20\pi\text{ rad}}{\frac{0 + 40\text{ rad/s}}{2}} = 3.1\text{ s}$$

5. $\omega_o = 3600$ rev/min $= 3.77 \times 10^2$ rad/s, $\theta = 50.0$ rev $= 3.14 \times 10^2$ rad, and $\omega = 0$.

$$\omega^2 = \omega_o^{\,2} + 2\alpha\theta .$$

$$0 = (3.77 \times 10^2\text{ rad/s})^2 + 2\alpha(3.14 \times 10^2\text{ rad})$$

$$\alpha = -2.26 \times 10^2\text{ rad/s}^2$$

6. $\omega = \omega_o + \alpha t = 2000\text{ rad/s} + (-80\text{ rad/s}^2)(10\text{ s}) = 1200\text{ rad/s}$

(b) $t = \frac{\omega - \omega_o}{\alpha} = \frac{0 - 2000\text{ rad/s}}{-80\text{ rad/s}^2} = 25\text{ s}$

7. (a) $v_t = r\omega$.

$80 \text{ m/s} = (200 \text{ m})\omega$

$\omega = 0.40 \text{ rad/s}$

(b) $a_t = \Delta v_t/\Delta t = 0$ because the tangential velocity is a constant.

8. (a) $a_t = \Delta v_t/\Delta t = \dfrac{95 \text{ m/s} - 80 \text{ m/s}}{10\text{s}} = 1.5 \text{ m/s}^2$

and $a_t = r\alpha$

$1.5 \text{ m/s}^2 = (200 \text{ m})\alpha$

$\alpha = 7.5 \times 10^{-3} \text{ rad/s}^2$

(b) $s = \overline{v}_t t = \dfrac{80 \text{ m/s} + 95 \text{ m/s}}{2}(10 \text{ s}) = 880 \text{ m}$

and $\theta = \dfrac{s}{r} = 880 \text{ m}/200 \text{ m} = 4.4 \text{ rad}$

9. $\omega = 200 \text{ rev/min} = 20.9 \text{ rad/s}$

$v = r\omega = (0.500 \text{ m})(20.9 \text{ rad/s}) = 10.5 \text{ m/s}$

$a_c = \dfrac{v^2}{r} = r\omega^2 = (0.500 \text{ m})(20.9 \text{ rad/s})^2 = 218 \text{ m/s}^2$

10. The deceleration of the rim of each tire is 2.00 m/s^2, and so the angular acceleration of each wheel is given by

$$\alpha = \frac{a}{r} = \frac{-2.00 \text{ m/s}^2}{0.300 \text{ m}} = -6.67 \text{ rad/s}^2$$

The initial angular speed ω_o of each wheel is $\omega_o = \dfrac{v}{r} = \dfrac{30.0 \text{ m/s}}{0.300 \text{ m}} = 100 \text{ rad/s}$.

Using $\omega^2 = {\omega_o}^2 + 2\alpha\theta$ and setting $\omega = 0$, we get

$$\theta = -\frac{{\omega_o}^2}{2\alpha} = \frac{(-100)^2}{2(-6.67)} = 750 \text{ rad} = 119 \text{ rev}$$

11. The mass of the 600 N student is 61.2 kg, and that of the 700 N student is 71.4 kg. Thus,

$$F = \frac{Gm_1 m_2}{r^2} = \frac{(6.67 \times 10^{-11} \text{ Nm}^2/\text{kg}^2)(61.2 \text{ kg})(71.4 \text{ kg})}{(0.500 \text{ m})^2} = 1.17 \times 10^{-6} \text{ N}$$

12. $g = \dfrac{GM}{R^2} = \dfrac{(6.67 \times 10^{-11} \text{ Nm}^2/\text{kg}^2)(5.98 \times 10^{24} \text{ kg})}{(4.66 \times 10^7 \text{ m})^2} = 0.184 \text{ m/s}^2$.

13. Since the force producing the centripetal acceleration of the satellite of mass m is the gravitational force exerted on it by Mars, mass M, we have

$$\frac{mv^2}{r} = \frac{GMm}{r^2}$$

From this, $v^2 = \frac{GM}{r}$. (1)

But we know that the velocity of the satellite is found from $v = \frac{2\pi r}{T}$. (2)

Eliminating v by substituting from (2) into (1) gives

$r^3 = \frac{GMT^2}{4\pi^2}$ For $M = 6.42 \times 10^{23}$ kg and $T = 459$ min $= 2.754 \times 10^4$ s, we find

$r = 9.37 \times 10^6$ m $= 5823$ mi

14. Since $\omega_1 = 8.0$ rev/s $= 16\pi$ rad/s, and $r_1 = 0.60$ m,

$v_1 = (0.60 \text{ m})(16\pi \text{ rad/s}) = 30.2 \text{ m/s}$

With the second sling, $\omega_2 = 6.0$ rev/s $= 12\pi$ rad/s, and $r_2 = 0.90$ m.

Thus, $v_2 = (0.90 \text{ m})(12\pi \text{ rad/s}) = 33.9 \text{ m/s}$.

(a) He obtains greater linear speed with the larger sling.

(b) $a_c = \frac{v^2}{r} = \frac{(30.2 \text{ m/s})^2}{0.60 \text{ m}} = 1.5 \times 10^3 \text{ m/s}^2$.

(c) $a_c = \frac{v^2}{r} = \frac{(33.9 \text{ m/s})^2}{0.90 \text{ m}} = 1.3 \times 10^3 \text{ m/s}^2$.

15. From $v_t = r\omega$ we have

$$\omega = \frac{v_t}{r} = \frac{1.2 \text{ m/s}}{0.20 \text{ m}} = 6.0 \text{ rad/s}$$

16. (a) $\omega = \omega_o + \alpha t$ gives

$\omega = 0 + (4.00 \text{ rad/s}^2)(2.00 \text{ s}) = 8.00 \text{ rad/s}$

(b) $v_t = r\omega. = (0.300 \text{ m})(8.00 \text{ rad/s}) = 2.40 \text{ m/s}$

$a_t = r\alpha = (0.300 \text{ m})(4.00 \text{ rad/s}^2) = 1.20 \text{ m/s}^2$

(c) $\theta = \theta_o + \omega_o t + \frac{1}{2}\alpha t^2$ yields

$\theta = 1.00 \text{ rad} + 0 + \frac{1}{2}(4.00 \text{ rad/s}^2)(2.00 \text{ s})^2 = 9.00 \text{ rad} = 516°$

(or at 156° counterclockwise from a horizontal reference line.)

8 ROTATIONAL EQUILIBRIUM AND ROTATIONAL DYNAMICS

PROBLEMS

*Indicates intermediate level problems.

1. A 45-kg, 5.0-m-long uniform ladder rests against a frictionless wall and makes an angle of 60° with a frictionless floor. Can an 80-kg person stand safely on the ladder, 2.0 m from the top, without causing the ladder to slip if a second person exerts a horizontal force of 500 N toward the wall at a point 3.5 m from the top of the ladder? (Note: All distances are measured along the ladder.)

2. The combination of an applied force and a frictional force produces a constant torque of 36.0 N · m on a wheel rotating about a fixed axis. The applied force acts for 6.00 s, during which time the angular speed of the wheel increases from 0 to 10.0 rad/s. The applied force is then removed, and the wheel comes to rest in 60.0 s. Find (a) the moment of inertia of the wheel, (b) the magnitude of the frictional torque, and (c) the total number of revolutions of the wheel. (38.)

3. A 48.0-kg diver stands at the end of a 3.00-m-long diving board. What torque does the weight of the diver produce about an axis perpendicular to and in the plane of the diving board through its midpoint?

4. Two children sit on a seesaw such that a 400-N child is 2.00 m from the support (the fulcrum). Where should a second child of weight 475 N sit in order to balance the system if the support is at the center of the plank?

5. The 400-N child of Problem 4 decides that she would like to seesaw alone. To do so, she moves the board such that its weight is no longer directly over the fulcrum. She finds that she will be balanced when she is 1.5 m to the left of the fulcrum and the center of the plank is 0.50 m to the right of the fulcrum. What is the weight of the plank?

*6. A dormitory door 2.50 m high and 1.00 m wide weighs 250 N, and its center of gravity is at its geometric center. The door is supported by hinges 0.250 m from top and bottom, each hinge carrying half the weight. Determine the horizontal component of the forces exerted by each hinge on the door.

7. A racing car has a mass of 1600 kg. The distance between the front and rear axles is 3 m. If the center of gravity of the car is 2 m from the rear axle, what is the normal force on each tire?

8. An iron trapdoor 1.25 m wide and 2.00 m long weighs 360 N and is hinged along the short dimension. Its center of gravity is at its geometric center. What force applied at right angles to the door is required to lift it (a) when it is horizontal and (b) when it has been opened so that it makes an angle of 30° with the horizontal? (Assume that the force is applied at the edge of the door opposite the hinges.)

9. A 4.50-kg ball on the end of a chain is whirled in a horizontal circle by an athlete. If the distance of the ball from the axis of rotation is 2.50 m, find the moment of inertia of the ball, assuming it can be considered as a point object.

10. What torque must the track star exert on the ball of Problem 9 to give it angular acceleration of 2.00 rad/s^2?

11. (a) Find the moment of inertia of a solid cylinder of mass 1.50 kg and radius 30.0 cm about an axis through its center. (b) Repeat for a solid sphere of the same mass and radius about an axis through its center.

12. The cylinder of Problem 11(a) is rotating at an angular velocity of 2.00 rev/s. What torque is required to stop it in 15.0 s?

13. An automobile tire, considered as a solid disk, has a radius of 35.0 cm and a mass of 6.00 kg. Find its rotational kinetic energy when rotating about an axis through its center at an angular velocity of 2.00 rev/s.

14. An automobile engine part is in the shape of a thin rod of mass 100 g and length 5.00 cm. When the rod is rotating at an angular velocity of 3.00 rad/s, find its kinetic energy when (a) rotating about an axis through a point 2.50 cm from each end. (b) Repeat when it is rotating about an axis through one end.

15. If the system of masses shown in Figure 8.2 is set into rotation about the x axis with an angular velocity of 2.5 rad/s, (a) find the kinetic energy of the system. (b) Repeat the calculation for the system in rotation at the same speed about the y axis.

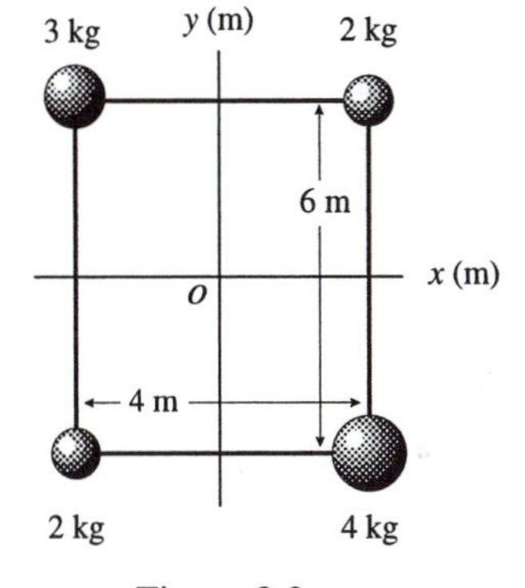

Figure 8.2

CHAPTER 8 SOLUTIONS

1. Since they are frictionless, the floor and wall exert only normal forces on the ladder. The normal force N_f, exerted on the ladder by the floor is found from $\Sigma F_y = N_f - w_{ladder} - w_{person} = 0$, which gives

$$N_f = w_{ladder} + w_{person} = 441 \text{ N} + 784 \text{ N} = 1225 \text{ N}$$

The sum of the torques acting on the ladder (choosing the pivot point at the top of the ladder) is

$$\Sigma\tau_{top} = (1225 \text{ N})(5.0 \text{ m})\cos 60° - (441 \text{ N})(2.5\text{m})\cos 60°$$

or $-(784 \text{ N})(2.0 \text{ m})\cos 60° - (1500 \text{ N})(3.5 \text{ m})\sin 60° = 212 \text{ N m} \neq 0.$

Since the net torque is not zero, the ladder is not in rotational equilibrium and will tend to slip.

2. (a) We first determine the angular acceleration due to the applied torque and the frictional torque by writing

$$\tau_{applied} + \tau_{friction} = I\alpha_1$$

with $\tau_{applied} + \tau_{friction} = 36.0 \text{ N m}$.

Now in six seconds, ω changes from 0 to 10.0 rad/s.

Using $\omega = \omega_o + \alpha_1 t$, we have $10.0 \text{ rad/s} = 0 + \alpha_1(6.00 \text{ s})$, giving

$$\alpha_1 = 1.67 \text{ rad/s}^2$$

Then, $36.0 \text{ N m} = (1.67 \text{ rad/s}^2)I$, or $I = 21.6 \text{ kg m}^2$.

(b) When the applied torque is removed, we have $\tau_{friction} = I\alpha_2$.

Using $\omega = \omega_o + \alpha_2 t$ we have $0 = 10.0 \text{ rad/s} + \alpha_2(60.0 \text{ s})$, so

$$\alpha_2 = -0.167 \text{ rad/s}^2$$

Therefore,

$$\tau_{friction} = (21.6 \text{ kg m}^2)(-0.167 \text{ rad/s}^2) = -3.60 \text{ N m}$$

and $|\tau_{friction}| = 3.60 \text{ N}$.

(c) The angular displacement during the first 6.00 s is

$\theta_1 = \omega_{1ave}(\Delta\tau) = \dfrac{0 + 10.0 \text{ rad/s}}{2}(6.00 \text{ s}) = 30.0 \text{ rad}$, and that during the last 60.0 s is

$$\theta_2 = \omega_{2ave}(\Delta\tau)_2 = \frac{10.0 \text{ rad/s}}{2 + 0}(60.0 \text{ s}) = 300 \text{ rad}$$

$$\theta = \theta_1 + \theta_2 = 30.0 \text{ rad} + 300 \text{ rad} = 330 \text{ rad} = 52.5 \text{ rev}$$

3. The weight of the diver is 470 N, and the torque about the pivot point through the center of the board is

$$\tau = (470\text{ N})(1.50\text{ m}) = 705\text{ N m (clockwise)}$$

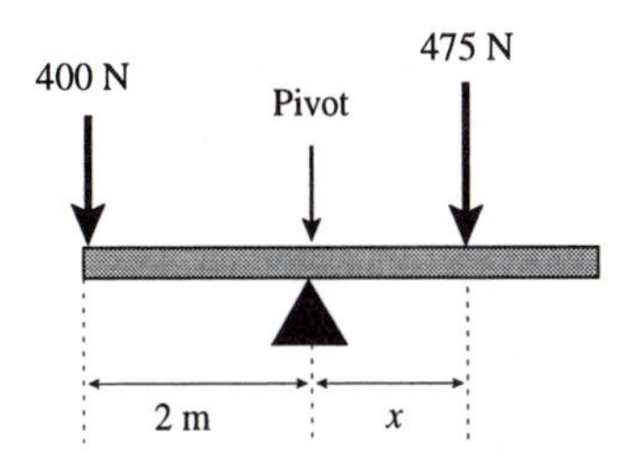

4. From $\Sigma\tau = 0$, we have

$$(400\text{ N})(2.00\text{ m}) - (475\text{ N})x = 0$$

$$x = 1.68\text{ m}$$

5. From $\Sigma\tau = 0$, we have

$$(400\text{ N})(1.5\text{ m}) - W(0.50\text{ m}) = 0$$

$$W = 1200\text{ N}$$

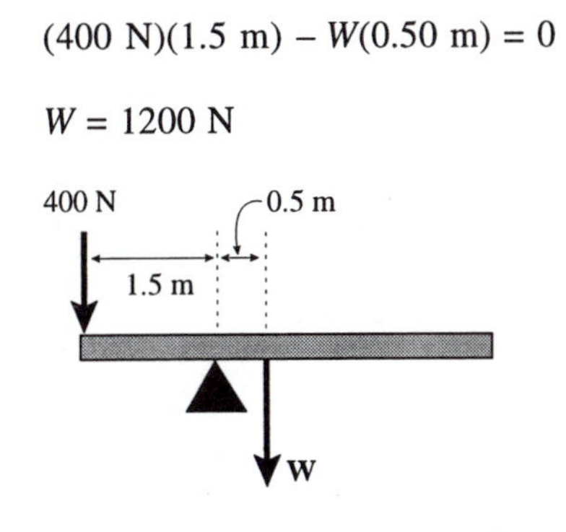

*6. We are given that $V_1 = V_2 = w/2 = 125$ N.

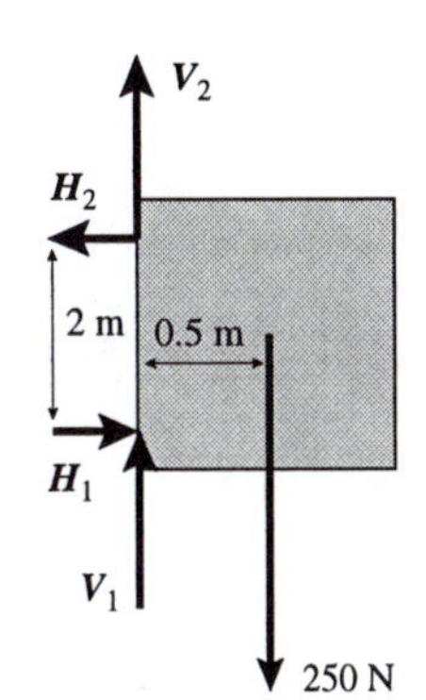

Take $\Sigma\tau = 0$ about indicated pivot pt.

$$-(250\text{ N})(0.500\text{ m}) + H_2(2.00\text{ m}) = 0$$

Giving, $H_2 = 62.5$ N

Now, use $\Sigma F_x = 0$.

$$H_1 - H_2 = 0$$

Therefore, $H_1 = 62.5$ N

The force on the lower hinge is a force of compression, and on the upper hinge the force is one of tension.

7. Choose the pivot point as shown, and $\Sigma\tau = 0$ gives

$$2F_R(0) - W(2 \text{ m}) + 2F_F(3 \text{ m}) = 0 \quad (1)$$

where $\mathbf{F_R}$ is the force on each rear tire and $\mathbf{F_F}$ is the force on each front tire. From (1),

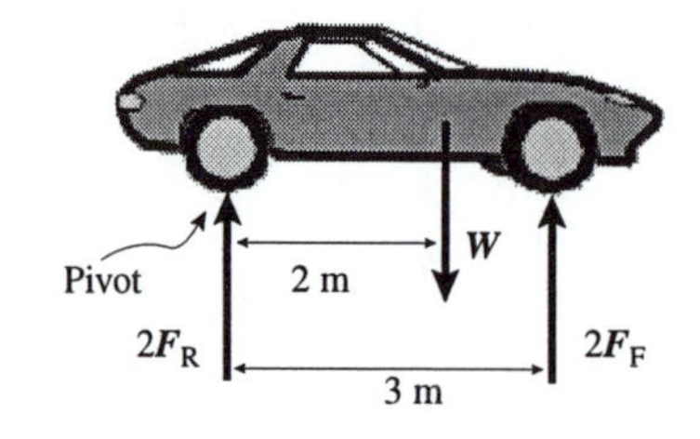

$$F_F = W/3$$

Now, use $\Sigma F_y = 0$.

$$2F_R - W - 2F_F = 0$$

Since $F_F = W/3$, we find $F_R = W/6$ (where $W = 15680$ N)

8. (a) When the door is horizontal, use $\Sigma\tau = 0$.

With the pivot point at the hinge, we have

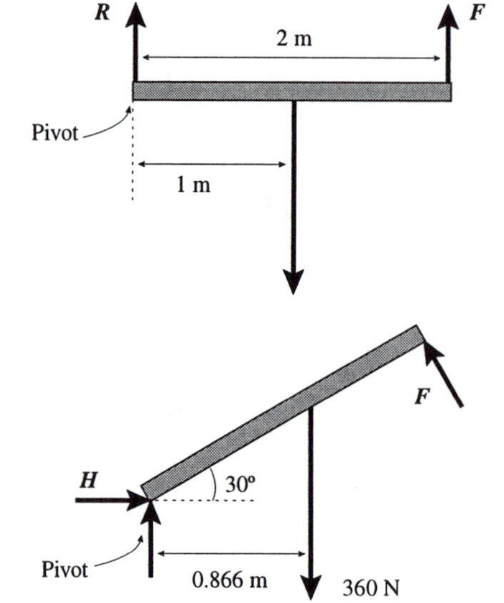

$$F(2.00 \text{ m}) - (360 \text{ N})(1.00 \text{ m}) = 0$$

or $F = 180$ N

(b) At 30.0° to the horizontal, and the pivot point at the hinge, $\Sigma\tau = 0$ gives

$$F(2.00 \text{ m}) - (360 \text{ N})(0.866 \text{ m}) = 0$$

where the lever arm for the 360 N force is

$$(1.00 \text{ m}) \cos 30° = 0.866 \text{ m}$$

We find, $F = 156$ N

9. $I = mr^2 = (4.50 \text{ kg})(2.50 \text{ m})^2 = 28.1 \text{ kg m}^2$.

10. $\tau = I\alpha = (28.1 \text{ kg m}^2)(2.00 \text{ rad/s}^2) = 56.3 \text{ N m}$

11. (a) For a cylinder $I = \frac{1}{2}MR^2$, where M is the mass of the cylinder and R the radius.

$$I = \frac{1}{2}(1.50 \text{ kg})(0.300 \text{ m})^2 = 6.75 \times 10^{-2} \text{ kg m}^2$$

(b) For a sphere, $I = \frac{2}{5}MR^2$, where M is the mass of the sphere and R the radius.

$$I = \frac{2}{5}(1.50 \text{ kg})(0.300 \text{ m}^2) = 5.40 \times 10^{-2} \text{ kg m}^2$$

12. $\omega_0 = 2.00$ rev/s $= 4\pi$ rad/s. Let us now find the angular acceleration of the cylinder.

    $\omega = \omega_0 + \alpha t$

    $0 = 4\pi \text{ rad/s} + \alpha(15.0 \text{ s})$

    $\alpha = -0.838 \text{ rad/s}^2$

    and $\tau = I\alpha$ gives $\tau = (6.75 \times 10^{-2} \text{ kg m}^2)(-0.838 \text{ rad/s}^2) = -5.65 \times 10^{-2}$ N m

13. (Note that 2.00 rev/s $= 4\pi$ rad/s.) Calculate the moment of inertia as

    $$I = \frac{1}{2}mr^2 = \frac{1}{2}(6.00 \text{ kg})(0.350 \text{ m})^2 = 3.68 \times 10^{-1} \text{ kg m}^2$$

    and $KE = \frac{1}{2}I\omega^2 = \frac{1}{2}(3.68 \times 10^{-1} \text{ kg m}^2)(4\pi \text{ rad/s})^2 = 29.0 \text{ J}$

14. (a) The moment of inertia of the rod when rotating about an axis through its center is

    $$I = \frac{1}{12}mL^2 = \frac{1}{12}(0.100 \text{ kg})(5.00 \times 10^{-2} \text{ m})^2 = 2.08 \times 10^{-5} \text{ kg m}^2$$

    and its kinetic energy is

    $$KE = \frac{1}{2}I\omega^2 = \frac{1}{2}(2.08 \times 10^{-5} \text{ kg m}^2)(3.00 \text{ rad/s})^2 = 9.38 \times 10^{-5} \text{ J}$$

    (b) When the rotation axis is at one end, we have

    $$I = \frac{1}{3}mL^2 = \frac{1}{3}(0.100 \text{ kg})(5.00 \times 10^{-2} \text{ m})^2 = 8.33 \times 10^{-5} \text{ kg m}^2$$

    and $KE = \frac{1}{2}I\omega^2 = \frac{1}{2}(8.33 \times 10^{-5} \text{ kg m}^2)(3.00 \text{ rad/s})^2 = 3.75 \times 10^{-4} \text{ J}$

15. (a) $I_x = 99 \text{ kg m}^2$

    $$KE = \frac{1}{2}I\omega^2 = \frac{1}{2}(99 \text{ kg m}^2)(2.5 \text{ rad/s})^2 = 310 \text{ J}$$

    (b) $I_y = 44 \text{ kg m}^2$

    $$KE = \frac{1}{2}I\omega^2 = \frac{1}{2}(44 \text{ kg m}^2)(2.5 \text{ rad/s})^2 = 140 \text{ J}$$

9 SOLIDS AND FLUIDS

PROBLEMS

*Indicates intermediate level problems.

1. A cylindrical aluminum pillar 7.00 high has a radius of 30.0 cm. If a 2000-kg sculpture is placed on top of the pillar, by how much is the pillar compressed?

2. A mass of 2.00 kg is supported by a copper wire of length 4.00 m and diameter 4.00 mm. Determine (a) the stress in the wire and (b) the elongation of the wire.

3. A uniform pressure of 5.00×10^4 Pa is exerted on a copper block having a volume of 10^{-3} m^3. What is the change in volume of the block?

4. During a tensile strength experiment, a small-diameter fiber is elongated by a force of 0.0250 N. The fiber has an initial length of 0.200 m and stretches a distance of 4.00×10^{24} m. Young's modulus for the material being tested is equal to 7.60×10^{10} Pa. Determine the diameter of the fiber.

5. A king orders a gold crown having a mass of 0.500 kg. When it arrives from the metalsmith, the volume of the crown is found to be 185 cm^3. Is the crown made of gold?

6. Determine the absolute pressure at the bottom of a lake that is 30.0 m deep.

7. If the column of mercury in a barometer stands at a height of 74.0 cm, what is the atmospheric pressure in Pa?

8. To what height in meters would the water in a water barometer stand on a day when a mercury barometer stands at 77.0 cm?

9. The pressure in a sealed water pipe is 2.50×10^5 Pa above atmospheric on the first floor of an apartment building. In comparison to this value, the pressure on the top floor is only 1.50×10^5 Pa. How tall is the building?

10. A popular television show has a feature in which various objects are compressed by a 90-ton hydraulic press, which has a cross-sectional area of about 400 cm^2. This force is generated by oil pushing against a smaller piston of area 2.00 cm^2. What force must the oil exert on this similar piston?

11. The U-shaped tube in Figure 9.1 contains mercury. (a) What is the absolute pressure on the left if the column height h = 20.0 cm? (b) What is the gauge pressure?

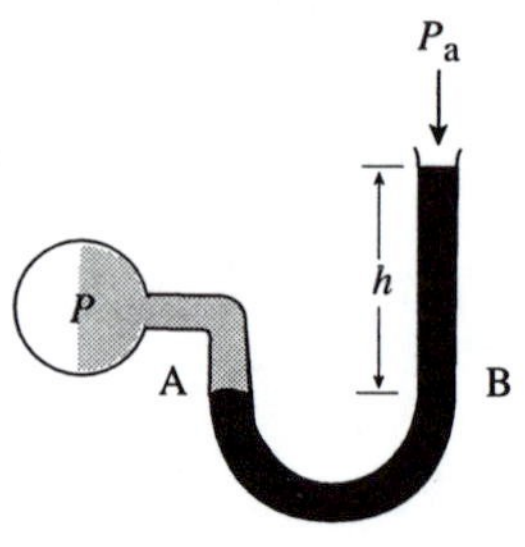

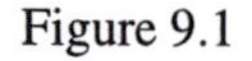
Figure 9.1

12. (a) Calculate the buoyant force on a solid object made of copper and having a volume of 0.200 m^3 if it is submerged in water. (b) What is the result if the object is made of steel?

*13. A solid object has a weight in air of 5.00 N. When it is suspended from a spring scale and submerged in water, the scale reads 3.50 N. What is the density of the object?

14. Careful analyses of photographs of a "sea monster" show that its volume is 52.0 m^3. Assume that the "sea monster" is weightless when immersed in seawater (density 1030 kg/m^3), and calculate its mass in metric tons (1 metric ton equals 1000 kg).

*15. An object of volume 450 cm^3 hangs from the end of a copper wire of cross-sectional area 1.50×10^{25} m^2. When the object is immersed in water the length of the wire decreases by 0.035 mm. What is the length of the wire?

CHAPTER 9 SOLUTIONS

1. $Y = \frac{\text{stress}}{\text{strain}}$, or strain $= \frac{\text{stress}}{Y}$, and $Y_{al} = 7.00 \times 10^{10}$ Pa

   so $\Delta L = \frac{FL}{AY} = \frac{(2.00 \times 10^3 \text{ kg})(9.80 \text{ m/s}^2)(7.00 \text{ m})}{\pi(0.300 \text{ m})^2(7.00 \times 10^{10} \text{ Pa})} = 6.93\ \mu\text{m}$

2. (a) Stress $= F/A = 19.6$ N/ 1.26×10^{-5} m^2 $= 1.56 \times 10^6$ Pa

   (b) Strain $= \Delta L/L =$ stress$/Y$

   Therefore,

   $\Delta L = L(\text{stress})/Y = (4.00 \text{ m})(1.56 \times 10^6 \text{ Pa})/11.0 \times 10^{10} \text{ Pa} = 5.67 \times 10^{-5}$ m

3. From the definition of bulk modulus, we find the change in volume to be

   $\Delta V = -(\Delta P)V/B = -(5.00 \times 10^4 \text{ Pa})(10^{-3} \text{ m}^3)/14.0 \times 10^{10} \text{ Pa}) = -3.57 \times 10^{-10} \text{ m}^3$.

4. From the definition of Young's modulus, the area is given by

   $A = (F)(L)/(\Delta L)(Y) = (0.0250 \text{ N})(2.00 \times 10^{-1} \text{ m})/(7.60 \times 10^{10} \text{ Pa})(4.00 \times 10^{-4} \text{ m})$

   $= 1.65 \times 10^{-10} \text{ m}^2$

   Also, $A = \frac{\pi d^2}{4}$ Hence, $d = 1.45 \times 10^{-5}$ m

5. $V = 185 \text{ cm}^3 = 1.85 \times 10^{-4} \text{ m}^3$

   $\rho = m/V = 5.00 \times 10^{-1} \text{ kg}/1.85 \times 10^{-4} \text{ m}^3 = 2.70 \times 10^3 \text{ kg/m}^3$ (crown is aluminum)

6. $P = P_a + \rho g h$

   $P = (1.013 \times 10^5 \text{ Pa}) + (10^3 \text{ kg/m}^3)(9.80 \text{ m/s}^2)(30.0 \text{ m}) = 3.95 \times 10^5$ Pa

7. $P = P_t + \rho g h$ where P_t = the pressure at the top of the column of mercury, which is small enough to be taken to be zero. Thus,

   $P = (13.6 \times 10^3 \text{ kg/m}^3)(9.80 \text{ m/s}^2)(0.740 \text{ m}) = 9.86 \times 10^4$ Pa

8. Atmospheric pressure for a barometer is found as $P = \rho g h$ regardless of the substance in the barometer. As a result,

   $\rho_{\text{water}} g h_{\text{water}} = \rho_{\text{mercury}} g h_{\text{mercury}}$

   or $h_{\text{water}} = \frac{13.6 \times 10^3}{1.00 \times 10^3}(0.770 \text{ m}) = 10.5$ m

9. We have the gauge pressure at the height of the first floor as

$$P_{g1} = \rho g h_1$$

and the gauge pressure at the top floor is

$$P_{g2} = \rho g h_2$$

Therefore,

$$P_{g2} - P_{g1} = \Delta P_g = \rho g \Delta h$$

or $\Delta h = \dfrac{\Delta P_g}{\rho g} = (1.50 \times 10^5 \text{ Pa} - 2.50 \times 10^5 \text{ Pa})/(10^3 \text{ kg/m}^3)(9.80 \text{ m/s}^2) = -10.2 \text{ m}$

Thus, the top floor is 10.2 m above the first floor (downward has been taken as the positive direction).

10. In this application, Pascal's law is written as

$$\frac{F_2}{A_2} = \frac{F_1}{A_1}$$

or $F_2 = \dfrac{A_2}{A_1} F_1 = \dfrac{2.00 \text{ cm}^2}{400 \text{ cm}^2} (1.80 \times 10^5 \text{ lbs}) = 900 \text{ lb}$

11. (a) $P = P_a + \rho g h = 1.013 \times 10^5 \text{ Pa} + (13.6 \times 10^3 \text{ kg/m}^3)(9.80 \text{ m/s}^2)(0.200 \text{ m}) = 1.28 \times 10^5 \text{ Pa}$.

(b) $P_g = P - P_a = 2.67 \times 10^4 \text{ Pa}$

12. (a) $B =$ (weight of water displaced) $= (\rho_w V_{\text{displaced}})g$

$$= (10^3 \text{ kg/m}^3)(0.200 \text{ m}^3)(9.80 \text{ m/s}^2) = 1.96 \times 10^3 \text{ N}$$

(b) The buoyant force remains the same because the same volume of water (and hence, weight) of water has been displaced.

13. The forces on the object when submerged are the tension in the string, the buoyant force of the water, and the weight of the object. We use $\Sigma F_y = 0$

$$B + T = w \qquad B + 3.50 \text{ N} = 5.00 \text{ N}$$

Therefore, $B = 1.50$ N (This is also the weight of the water displaced.)

We also know,

$$V_{\text{displaced water}} = V_{\text{object}}$$

Thus, the buoyant force may be written as

$$B = (\rho_w V_{\text{object}})g$$

or, $V_{\text{object}} = B/\rho_w g$ (1)

13. (continued)
The density of the object is found as

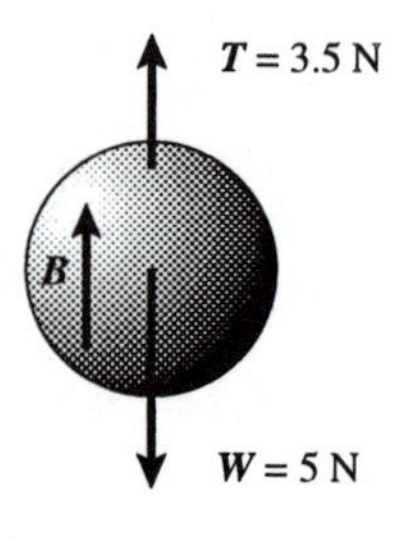

$$\rho = m_{\text{object}}/V_{\text{object}} \qquad (2)$$

We eliminate the volume of the object from equation (2) by use of (1) to find,

$$\rho = \frac{(w_{\text{object}})(\rho_w)}{B} = \frac{(5.00\ \text{N})(10^3\ \text{kg/m}^3)}{1.50\ \text{N}} = 3.33 \times 10^3\ \text{kg/m}^3$$

14. Because the sea monster floats, its weight, w, is equal to the buoyant force (the weight of water displaced by the 52.0 m^3 volume of the monster). We have

$$w = mg = \rho_w Vg$$

or, $m = \rho_w V = (1.03 \times 10^3\ \text{kg/m}^3)(52.0\ \text{m}^3) = 5.36 \times 10^4\ \text{kg} = 53.6$ metric tons

15. When in water the buoyant force is

$$B_w = \rho_w Vg = (10^3\ \text{kg/m}^3)(450 \times 10^{-6}\ \text{m}^3)(9.80\ \text{m/s}^2) = 4.41\ \text{N}$$

When the object is in air, this buoyant force will no longer be present and an additional weight of 4.41 N will have to be supported by the wire. This much additional force stretches the wire by 0.0350 mm. From the definition of Young's modulus, we find

$$L = \frac{(\Delta L)(A)(Y)}{F} = \frac{(3.50 \times 10^{-5}\ \text{m})(1.50 \times 10^{-5}\ \text{m}^2)(11.0 \times 10^{10}\ \text{Pa})}{4.41\ \text{N}} = 13.1\ \text{m}$$

10 THERMAL PHYSICS

PROBLEMS

*Indicates intermediate level problems.

1. The temperature in the interior of some stars is approximately 2.00×10^7 K. Express this temperature in (a) degrees Fahrenheit and (b) degrees Celsius.

2. Recently the temperature on a certain day changed by 54° F. Express this change in temperature on (a) the Celsius scale and (b) the Kelvin scale.

3. A copper steam pipe is 2.0000 m long and is installed in a basement when the temperature is 20°C. What is the length of the pipe when it carries steam at 120°C?

4. A structural steel beam is 20.0 m long when installed at 20°C. How much will its length change over the temperature extremes of 26°C to 30°C?

5. The concrete sections of a certain highway are designed to have a length of 30.0 m. The sections are poured and cured at 10°C. What minimum spacing should the engineer leave between the sections to eliminate buckling if the concrete is to reach a temperature of 45°C?

*6. A copper pipe has a length of 2.00 m when the temperature is 20°C. If it is rigidly clamped in place at 20°C, calculate the thermal stress set up in the pipe when the temperature is raised to 100°C.

*7. At what temperature would the rms speed of helium atoms (mass = 6.66×10^{-27} kg) equal (a) the escape velocity from Earth, 1.12×10^4 m/s? (b) the escape velocity from the Moon, 2.37×10^3 m/s?

8. An ideal gas is held in a container at constant volume. Initially, its temperature is 20°C and its pressure is 3.00 atm. Find the pressure when its temperature is increased to 50°C.

9. The pressure of a gas in a container is tripled while its volume is halved. What is the ratio of the final to the original temperature of the gas?

10. (a) A 10.0-liter tank is to be filled with oxygen at 20°C to a pressure of 50.0 atm. Assume the ideal gas equation applies at this high pressure and find the mass of oxygen required. (b) Repeat assuming the gas is helium.

*11. (a) Use an ideal gas equation to find the density of oxygen at standard temperature and pressure. (b) Repeat for helium.

12. (a) Find the number of molecules in 1.00 m^3 of air at atmospheric pressure and 27°C. (b) Repeat Part (a) under the condition at which the pressure has been reduced to 10^{-11} Pa at 27°C.

13. (a) Calculate the rms speed of an H_2 molecule when the temperature is 100°C. (b) Repeat the same calculation for an N_2 molecule.

14. What is the temperature at which the rms speed of nitrogen molecules equals the rms speed of helium at 20°C?

15. How many kilograms of nitrogen are contained in a tank whose volume is 0.750 m^3 when the gauge pressure is 100 atm and the temperature is 27°C?

CHAPTER 10 SOLUTIONS

1. To illustrate the approach, we are going to assume here (incorrectly) that the temperature of the star is known to 8 significant figures. The Celsius temperature is

   $$T_C = T_K - 273 = 19{,}999{,}727°C$$

   and, from, $T_F = \frac{9}{5}\,T_C + 32$, we find $T_F = 35{,}999{,}954°F$

2. (a) We use

   $$\Delta T_C = \frac{5}{9}\,(\Delta T_F)$$

   which gives $\Delta T_C = \frac{5}{9}\,(54\ °F) = 30°C$

   (b) The relationship between a change on the kelvin scale and the Celsius scale is

   $$\Delta T_K = \Delta T_C$$

   Thus, $\Delta T_K = 30\ K$

3. $\Delta L = \alpha L_o \Delta T = (17 \times 10^{-6}\ °C^{-1})(2.0000\ m)(100°C) = 3.4\ mm$

   Thus, $L = L_o + \Delta L = 2.0000\ m + .0034\ m = 2.0034\ m$

4. The temperature range is $\Delta T = 30°C - (-6°C) = 36°C$, and the change in length is given by,

   $$\Delta L = \alpha L_o \Delta T = (11.0 \times 10^{-6}\ °C^{-1})(20.0\ m)(36°C) = 7.92\ mm$$

5. The gap should be equal to the amount of expansion of one section. Thus,

   $$\Delta L = \alpha L_o \Delta T = (12 \times 10^{-6}\ °C^{-1})(30.0\ m)(35.0°C) = 1.26 \times 10^{-2}\ m = 1.26\ cm$$

6. As the temperature of the pipe is raised, its length increases by

   $$\Delta L = \alpha L_o \Delta T = (17 \times 10^{-6}\ °C^{-1})(2.00\ m)(80.0\ °C) = 2.72 \times 10^{-3}\ m$$

   Thus, the strain created is

   $$\text{Strain} = \frac{\Delta L}{L_o} = \frac{2.72 \times 10^{-3}\ m}{2.00\ m} = 1.36 \times 10^{-3}$$

   and the stress is found from the definition of Young's modulus.

   $$\text{Stress} = Y(\text{strain}) = (11 \times 10^{10}\ Pa)(1.36 \times 10^{-3}) = 1.50 \times 10^{8}\ Pa$$

7. (a) From $v_{rms} = \sqrt{\dfrac{3kT}{m}}$, we find the temperature as:

$$T = \frac{(6.66 \times 10^{-27}\ \text{kg})(1.12 \times 10^{4}\ \text{m/s})^2}{3(1.38 \times 10^{-23}\ \text{J/mol K})} = 2.02 \times 10^{4}\ \text{K}.$$

(b) $$T = \frac{(6.66 \times 10^{-27}\ \text{kg})(2.37 \times 10^{3}\ \text{m/s})^2}{3(1.38 \times 10^{-23}\ \text{J/mol K})} = 9.04 \times 10^{2}\ \text{K}.$$

8. We may use of the ideal gas equation expressed as

$$\frac{P_F V_F}{T_F} = \frac{P_i V_i}{T_i}$$

Because the container is kept at constant volume, we have

$$\frac{P_F}{T_F} = \frac{P_i}{T_i}$$

or, $$\frac{P_F}{323\ \text{K}} = \frac{3.00\ \text{atm}}{293\ \text{K}}$$

and $P_F = 3.31$ atm

9. From $\dfrac{P_F V_F}{T_F} = \dfrac{P_i V_i}{T_i}$, we have

$$\frac{T_F}{T_i} = \frac{P_F V_F}{P_i V_i} = \frac{(3P_i)(V_i/2)}{P_i V_i} = \frac{3}{2}$$

10. We first find the number of moles of gas required as

$$n = \frac{PV}{RT} = \frac{(5.07 \times 10^{6}\ \text{Pa})(10^{-2}\ \text{m}^3)}{(8.31\ \text{J/mol K})(293\ \text{K})} = 20.8\ \text{mol}$$

But, $n = \dfrac{m}{M}$, so $m = nM$

(a) If the gas is oxygen, $M = 32$ g/mol (a diatomic gas)

Thus, $m = (20.8\ \text{mol})(32\ \text{g/mol}) = 666$ g

(b) If the gas is helium, $M = 4$ g/mol

and, $m = (20.8\ \text{mol})(4\ \text{g/mol}) = 83.2$ g

11. (a) To find the density, we use $\rho = \dfrac{PM}{RT}$

$$\rho = \frac{(1.013 \times 10^{5}\text{ Pa})(32 \times 10^{-3}\text{ kg/mol})}{(8.314\text{ J/mol K})(273\text{ K})} = 1.43\text{ kg/m}^3$$

(b) The approach is the same here as in part (a), with $M = 4 \times 10^{-3}$ kg/mol for helium. We find,

$$\rho = 1.79 \times 10^{-1}\text{ kg/m}^3$$

12. We use

$$PV = NkT$$

$$N = P\,\frac{1.00\text{ m}^3}{(1.38 \times 10^{-23}\text{ N m/mol K})(300\text{ K})} = P(2.42 \times 10^{20}\text{ mol m}^2\text{/N}) \quad (1)$$

(a) At atmospheric pressure, (1) gives

$$N = (1.013 \times 10^{5}\text{ Pa})(2.42 \times 10^{20}\text{ mol m}^2\text{/N}) = 2.45 \times 10^{25}\text{ molecules}$$

(b) Using the same approach as in (a) for a pressure of 10^{-11} Pa gives

$$N = 2.42 \times 10^{9}\text{ molecules}$$

13. (a) The mass, m, of a molecule is given by m = molecular weight/N_a. For diatomic hydrogen, we have

$$m = (2 \times 10^{-3}\text{ kg/mol})/(6.02 \times 10^{23}\text{ molecules/mol}) = 3.32 \times 10^{-27}\text{ kg/mol}$$

Thus,

$$v_{\text{rms}} = \sqrt{\frac{3kT}{m}} = \sqrt{\frac{3(1.38 \times 10^{-23}\text{ J/mol K})(373\text{ K})}{3.32 \times 10^{-27}\text{ kg/mol}}} = 2.16 \times 10^{3}\text{ m/s}$$

(b) We use the same technique as in part (a). The mass of a nitrogen molecule is found to be 4.65×10^{-26} kg/molecule, and the rms speed is 576 m/s.

14. $\langle KE \rangle = \dfrac{1}{2} m v^2{}_{\text{rms}} = \dfrac{3}{2} kT$

so $T = \dfrac{m}{3k} v^2{}_{\text{rms}}$

For helium

$$m = 6.64 \text{ x } 10^{-27}\text{ kg}$$

Thus, at 20°C,

$$v^2{}_{\text{rms}} = \frac{3kT}{m} = \frac{3(1.38 \times 10^{-23}\text{ J/mol K})(293\text{ K})}{6.64 \times 10^{-27}\text{ kg/mole}} = 1.81 \times 10^{6}\text{ m}^2/\text{s}^2$$

14. (continued)
For nitrogen molecules,

$$m = \frac{28 \text{ g/mole}}{6.02 \times 10^{23} \text{ molecules/mol}} = 4.65 \times 10^{-23} \text{ g} = 4.65 \times 10^{-26} \text{ kg}$$

Therefore, when $v^2_{\text{rms}} = 1.81 \times 10^6 \text{ m}^2/\text{s}^2$ (as it is for helium at 20°C)

$$T = \frac{(4.65 \times 10^{-26} \text{ kg/mol})}{3(1.38 \times 10^{-23} \text{ J/mol K})}(1.81 \times 10^6 \text{ m}^2/\text{s}^2) = 2040 \text{ K or } 1770°\text{C}$$

15. Note that the absolute pressure is 101 atm (1.02×10^7 Pa. From the ideal gas equation, we have

$$n = \frac{PV}{RT} = \frac{(1.02 \times 10^7 \text{ Pa})(0.750 \text{ m}^3)}{(8.31 \text{ J/mol K})(300 \text{ K})} = 3.08 \times 10^3 \text{ mol}$$

Nitrogen has a molecular weight of 28 g/mol. Thus, the mass of the gas is

$$m = nM = (3.08 \times 10^3 \text{ mol})(28 \times 10^{-3} \text{ kg/mol}) = 86.2 \text{ kg}$$

11 HEAT

PROBLEMS

*Indicates intermediate level problems.

1. Consider Joule's apparatus described in Figure 11.1. The two masses are 1.50 kg each, and the tank is filled with 200 g of water. What is the increase in temperature of the water after the masses fall through a distance of 3.00 m?

2. A construction worker drops a hot 100-g iron rivet at 500°C into a bucket containing 500 g of mercury at 20°C. Assuming that no heat is lost to the surroundings or the bucket, what is the final temperature of the rivet and mercury?

3. An unknown liquid of mass 400 g at a temperature of 80°C is poured into 400 g of water at 40°C. The final equilibrium temperature of the mixture is 49°C. What is the specific heat of the unknown liquid?

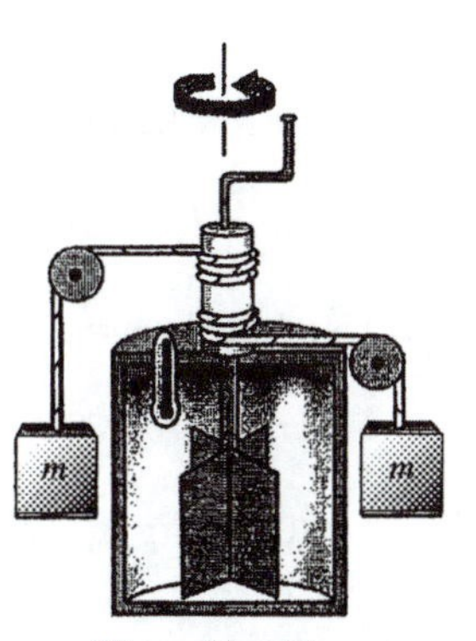

Figure 11.1

*4. In a foundry operation, molten lead of mass 90.0 g at 327.3°C is poured into an iron casting block that has a mass of 300 g and is initially at 20.0°C. What is the equilibrium temperature of the system? Assume no heat losses to the surroundings.

*5. A Styrofoam cup contains 200 g of mercury at 0.0°C. To this is added 50.0 g of ethyl alcohol at 50.0°C and 100 g of water at 100°C. (a) What is the final temperature of the mixture? (b) How much heat was gained or lost by (i) the mercury, (ii) the alcohol, and (iii) the water? (The specific heat of ethyl alcohol is 2430 J/kg · C° and the heat capacity of the Styrofoam cup is negligible.)

*6. Determine the final state when 20.0 g of 0.00°C ice and 10.0 g of 100°C steam are mixed together in an insulated container.

*7. A 300-g ice cube at 0.00°C is placed in 200 g of water at 10.0°C. The water is in a 100-g aluminum container. How much ice melts?

*8. A glass window pane has an area of 3.0 m^2 and a thickness of 0.60 cm. If the temperature difference between its faces is 25°C, how much heat flows through the window per hour?

CHAPTER 11 SOLUTIONS

1. $Q = \Delta PE = mgh = 2(1.50 \text{ kg})(9.80 \text{ m/s}^2)(3.00 \text{ m}) = 88.2 \text{ J}$

   Also, $\Delta T = \dfrac{Q}{mc} = \dfrac{88.2 \text{ J}}{(0.2 \text{ kg})(4.186 \times 10^3 \text{ J/kg °C})} = 0.105°\text{C}$

2. Use heat loss = heat gain: $m_{\text{rivet}} c_{\text{rivet}} \Delta T_{\text{rivet}} = m_{\text{Hg}} c_{\text{Hg}} \Delta T_{\text{Hg}}$, or

   $(0.100 \text{ kg})(448 \text{ J/kg°C})(500 - T_f) = (0.500 \text{ kg})(138 \text{ J/kg °C})(T_f - 20°\text{C})$

   yielding $T_f = 210°\text{C}$.

3. Use heat loss = heat gain

   $m_{\text{unknown}} c_u \Delta T_u = m_{\text{water}} c_w \Delta T_w$

   $(0.400 \text{ kg})c_u(31°\text{C}) = (0.400 \text{ kg})(4186 \text{ J/kg °C})(9°\text{C})$

   $c_u = 1220 \text{ J/kg °C}$

4. heat loss = heat gain

   $(0.0900 \text{ kg})(2.45 \times 10^4 \text{ J/kg}) + (0.0900 \text{ kg})(128 \text{ J/kg °C})(327.3°\text{C} - T_f)$

   $= (0.300 \text{ kg})(448 \text{ J/kg°C})(T_f - 20.0°\text{C})$

   From which, $T_f = 59.4°\text{C}$

5. $\Delta Q_{\text{system}} = 0$

   $= (0.200 \text{ kg})(138\text{J/kg°C})(T_f - 0°\text{C}) + (0.0500 \text{ kg})(2430 \text{ J/kg°C})(T_f - 50.0°\text{C})$

   $+ (0.100 \text{ kg})(4186\text{J/kg°C})(T_f - 100°\text{C})$

   Gives $T_f = 84.4°\text{C}$

   (b) $\Delta Q_{\text{mercury}} = (0.200 \text{ kg})(138\text{J/kg°C})(84.4°\text{C} - 0°\text{C}) = 2330 \text{ J}$

   $\Delta Q_{\text{alc}} = (0.0500 \text{ kg})(2430\text{J/kg°C})(84.4°\text{C} - 50.0°\text{C}) = 4184 \text{ J}$

   $\Delta Q_{\text{water}} = (0.100 \text{ kg})(4186\text{J/kg°C})(84.4°\text{C} - 100°\text{C}) = -6514 \text{ J}$

*6. Heat required to melt ice = $(0.0200 \text{ kg})(3.34 \times 10^5 \text{ J/kg}) = 6680 \text{ J}$

   Heat steam can give up in condensing = $(0.0100 \text{ kg})(2.26 \times 10^6 \text{ J/kg}) = 22{,}600 \text{ J}$

   Therefore, the steam can melt all the ice and still be able to yield as much as 22,600 J – 6680 J = 15,920 J to raise the temperature of the melted ice. The heat required to raise the melted ice to 100°C

   $= (0.0200 \text{ kg})(4186 \text{ J/kg°C})(100°\text{C}) = 8370 \text{ J}$

6. (continued)
Thus, the ice can absorb 6680 J + 8370 J = 15,050 J from the steam before it reaches the same temperature as the steam (i.e. thermal equilibrium).

So, we condense $\dfrac{15050 \text{ kg}}{2.26 \times 10^6 \text{ J/kg}} = 6.66$ g of steam, and we are left with a mixture of 3.34 g of steam and 26.66 g of liquid water all at 100°C.

*7. The final temperature is 0.00°C since not all of the ice can be melted.

heat gain = heat loss

(heat to melt ice) = (heat loss as water cools to 0.00°C) + (heat loss as cup cools to 0.00°C)

$$m(3.33 \times 10^5 \text{ J/kg}) = (0.200 \text{ kg})(4186 \text{ J/kg °C})(10.0\text{°C}) + (0.100 \text{ kg})(900 \text{ J/kg °C})(10.0\text{°C})$$

$$m = 2.78 \times 10^{-2} \text{ kg} = 27.8 \text{ g}$$

8. $$H = \frac{\Delta Q}{\Delta t} = \frac{kA\Delta T}{L} = \frac{(0.8 \text{ J/s m C})(3.0 \text{ m}^2)(25\text{°C})}{6.0 \times 10^{-3} \text{ m}} = 10^4 \text{ J/s}$$

Thus, in one hour, $\Delta Q = Ht = (10^4 \text{ J/s})(3600 \text{ s}) = 3.6 \times 10^7 \text{ J} = 8.6 \times 10^6$ cal.

12 THE LAWS OF THERMODYNAMICS

PROBLEMS

*Indicates intermediate level problems.

*1. A sample of gas is compressed to one-half its initial volume at a constant pressure of 1.25×10^5 Pa. During the compression, 100 J of work is done on the gas. Determine the final volume of the gas.

2. A cylinder of gas is compressed by a piston from an initial volume of 125 liters to a final volume of 90.0 liters. The compression occurs at constant pressure, and the work done on the gas by the piston is 10^4 J. What is the gas pressure during the compression?

3. A thermodynamic system undergoes a process in which its internal energy decreases by 500 J. If at the same time, 220 J of work is done on the system, find the heat transferred to or from the system.

*4. A person takes in a breath at 0.00°C and holds its until the air warms to 37.0°C. The air taken into the lungs has an initial volume of 0.600 L and a mass equal to 7.70×10^{-4} kg. (a) Determine the work done on the lungs by the air if the pressure remains constant at 1.00 atm. (b) Determine the heat added to the air. (c) Determine the change of internal energy of the air. (Use $c = 1010$ J/kg · °C for air.)

*5. One mole of an ideal gas does 3000 J of work on the surroundings as it expands isothermally to a final pressure of 1.00 atm and volume of 25.0 L. Determine (a) the initial volume and (b) the temperature of the gas.

6. In each cycle, a heat engine absorbs 375 J of heat and performs 25.0 J of work. Find (a) the efficiency of the engine and (b) the heat expelled in each cycle.

7. The heat absorbed by an engine is three times greater than the work it performs. (a) What is the thermal efficiency? (b) What fraction of the heat absorbed is expelled to the cold reservoir?

8. The work done during an isothermal expansion of an ideal gas from an initial volume V_i to a final volume V_f is

$$W = nRT \ln\left(\frac{V_f}{V_i}\right)$$

Calculate the work done by one mole of an ideal gas at 300°C as it expands isothermally until its volume is tripled.

9. An ice tray contains 500 g of water at 0.00°C. Calculate the entropy change of the water as it freezes completely and slowly at 0.00°C.

10. One mole of an ideal gas at 1.0 atm pressure and at 300°C expands isothermally until its volume is tripled. It then is compressed to its original volume at constant pressure. Sketch these two processes on a *PV* diagram.

*11. An ideal refrigerator (or heat pump) is equivalent to a Carnot engine running in reverse. That is, heat Q_c is absorbed from a cold reservoir and heat Q_h is rejected to a hot reservoir. Show that the work that must be supplied to run the refrigerator is given by

$$W = \frac{T_h - T_c}{T_c} Q_c$$

CHAPTER 12 SOLUTIONS

1. $W = -P(V_f - V_i)$ (W is negative because work is done on the gas.)

   $W = -P(V_f - 2V_f) = PV_f$

   Thus, $100 \text{ J} = (1.25 \times 10^5 \text{ Pa})V_f$

   We find, $V_f = 8.00 \times 10^{-4} \text{ m}^3 = 800 \text{ cm}^3$.

2. The work is done on the gas, so

   $W = -P\Delta V$

   Thus, $P = \dfrac{-W}{\Delta V} = \dfrac{-10^4 \text{ J}}{-35.0 \times 10^{-3} \text{ m}^3} = 2.86 \times 10^5 \text{ Pa}$

3. $\Delta U = Q - W$ or $Q = \Delta U + W = -500\text{J} - 220 \text{ J} = -720 \text{ J}$

4. The initial and final pressures are 1 atm, and we can treat the system as an ideal gas, $PV = nRT$. We have

   $\dfrac{V_f}{T_f} = \dfrac{V_i}{T_i}$ or $V_f = \dfrac{T_f}{T_i} V_i = \dfrac{310 \text{ K}}{273 \text{ K}}(0.600 \times 10^{-3} \text{ m}^3) = 6.81 \times 10^{-4} \text{ m}^3$

   and $V_i = 6.00 \times 10^{-4} \text{ m}^3$

   so $\Delta V = 8.10 \times 10^{-5} \text{ m}^3$

   (a) $W = P\Delta V = (1.013 \times 10^5 \text{ Pa})(8.10 \times 10^{-5} \text{ m}^3) = 8.24 \text{ J}$

   (b) $Q = mc\Delta T = (7.70 \times 10^{-4} \text{ kg})(1010 \text{ J/kg°C})(37.0\text{°C}) = 28.8 \text{ J}$

   (c) $\Delta U = 28.8 \text{ J} - 8.24 \text{ J} = 20.5 \text{ J}$

5. (b) From the ideal gas equation,

   $T = \dfrac{P_f V_f}{nR} = \dfrac{(1.013 \times 10^5 \text{ Pa})(25.0 \times 10^{-3} \text{ m}^3)}{(1 \text{ mol})(8.31 \text{ J/mol K})} = 305 \text{ K} = 32.0\text{°C}$

   (a) $\ln\dfrac{V_f}{V_i} = \dfrac{W}{nRT} = \dfrac{3000 \text{ J}}{(1 \text{ mol})(8.31 \text{ J/mol K})(305 \text{ K})} = 1.19$

   $\dfrac{V_f}{V_i} = 3.27$ and $V_i = \dfrac{V_f}{3.27} = \dfrac{25.0 \text{ L}}{3.27} = 7.65$ liters

6. (a) $Eff = \dfrac{W}{Q_h} = \dfrac{25.0 \text{ J}}{375 \text{ J}} = 0.067$

   (b) $Q_c = Q_h - W = 375 \text{ J} - 25.0 \text{ J} = 350 \text{ J}$

7. (a) $Eff = \frac{W}{Q_h} = \frac{W}{3W} = \frac{1}{3} = 0.333$

(b) $Q_c = Q_h - W = 3W - W = 2W$

Therefore, $\frac{Q_c}{Q_h} = \frac{2W}{3W} = \frac{2}{3}$

8. We are given that: $W = nRT \ln\left(\frac{V_f}{V_i}\right)$, with

$$n = 1 \text{ mole}, \quad \frac{V_f}{V_i} = 3 \quad \text{and} \quad T = 300 + 273 = 573 \text{ K}$$

Therefore, $W = (1 \text{ mol})\left(8.31 \frac{\text{J}}{\text{mol K}}\right)(573 \text{ K}) \ln(3) = 5232$ J.

9. The heat lost by the water is

$$Q = -mL_f = -(0.500 \text{ kg})(3.33 \times 10^5 \text{ J/kg}) = -1.67 \times 10^5 \text{ J}$$

Therefore, $\Delta S = \frac{Q}{T} = \frac{-1.67 \times 10^5 \text{ J}}{273 \text{ K}} = -612$ J/K

10. From the ideal gas law, the final pressure of the gas is

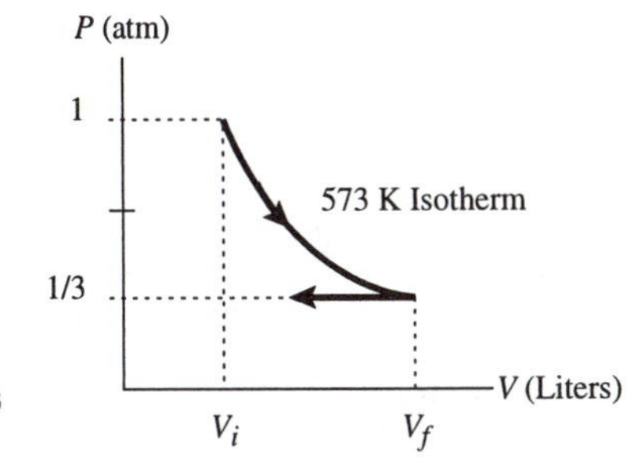

$$P_f = P_i\frac{V_i}{V_f} = \frac{1}{3}P_i = \frac{1}{3} \text{ atm}$$

and the initial volume is

$$V_i = \frac{nRT_i}{P_i} = \frac{(1\text{mol})(8.31 \text{ J/mol K})(573 \text{ K})}{1.013 \times 10^5 \text{ Pa}} = 4.7 \times 10^{-2} \text{ m}^3$$

The sketch for this process is shown at the right.

11. (a) For a complete cycle, $\Delta U = 0$ and $W = Q_h - Q_c = Q_c\left[\frac{Q_h}{Q_c} - 1\right]$

We have already shown that for a Carnot cycle (and only for a Carnot cycle) that $\frac{Q_h}{Q_c} = \frac{T_h}{T_c}$

Therefore, $W = Q_c\left(\frac{T_h - T_c}{T_c}\right)$

13 VIBRATIONS AND WAVES

PROBLEMS

1. (a) A mass of 400 g is suspended from a spring hanging vertically, and the spring is found to stretch 8.00 cm. Find the spring constant. (b) How much will the spring stretch if the suspended mass is 575 g?

2. A 3.00-kg mass is attached to a spring and pulled out horizontally to a maximum displacement from equilibrium of 0.500 m. What spring constant must the spring have if the mass is to achieve an acceleration equal to that of gravity?

3. When a 4.00-kg mass is hung vertically on a certain light spring that obeys Hooke's law, the spring stretches 2.50 cm. If the 4.00-kg mass is removed, (a) how far will the spring stretch if a 1.50-kg mass is hung on it? (b) How much work must an external agent do to stretch the same spring 4.00 cm from its unstretched position?

4. A spring of spring constant 19.6 N/m is compressed 5.00 cm. A mass of 0.300 kg is attached to the spring and released from rest. Find (a) the maximum elastic potential energy stored in the spring and (b) the maximum speed of the mass.

5. The mat of a trampoline is held by 32 springs, each having a spring constant of 5000 N/m. A 40.0-kg person jumps from a 1.93-m-high platform onto the trampoline. Determine the stretch of each of the springs. Assume that the springs were initially unstretched and that they stretch equally.

6. In an arcade game a 0.100-kg disk is shot across a frictionless horizontal surface by compressing it against a spring and releasing it. If the spring has a spring constant of 200 N/m and is compressed from its equilibrium position by 6.00 cm, find the speed with which the disk slides across the surface.

7. A telephone cord is 4.0 m long. The cord has a mass of 0.20 kg. If a transverse wave pulse travels from the receiver to the telephone box in 0.10 s, what is the tension in the cord?

8. A 0.500-kg object is attached to a spring of spring constant 30.0 N/m and released from rest at $t = 0$ from a position 0.250 m from the equilibrium. Find the equation that describes its x location as a function of time.

9. The period of a 50.0-cm-long pendulum is found to be 1.40 s. What is the acceleration due to gravity at this location?

10. A pendulum is to be used in a clock. What length should the pendulum be so that its period of vibration is 1.00 s?

11. A cork resting on the surface of a pond bobs up and down two times per second on some ripples having a wavelength of 8.50 cm. If the cork is 10.0 m from shore, how long does it take a ripple passing the cork to reach the shore?

12. How far in meters does light travel in one year? This distance is called a light-year.

13. A phone cord is 4.00 m long. The cord has a mass of 0.200 kg. If a transverse wave pulse travels from the receiver to the phone box in 0.100 s, what is the tension in the cord?

14. One end of a string 3.00 m long is attached to a wall while the other end hangs over a pulley and is attached to a hanging 2.00-kg mass. The speed of a pulse on the string is observed to be 15.0 m/s. What is the mass of the string?

15. A string under a tension of 100 N has a wave on it traveling at 90.0 m/s. What is the speed of the wave on this string if the tension is doubled?

16. The amplitude of a system moving with simple harmonic motion is doubled. Determine the changes in (a) total energy, (b) maximum velocity, (c) maximum acceleration, and (d) period.

CHAPTER 13 SOLUTIONS

1. (a) At equilibrium, $mg = kx$. Thus,

$$k = \frac{mg}{x} = \frac{(0.400\text{ kg})(9.80\text{ m/s}^2)}{(8.00 \times 10^{-2}\text{ m})} = 49.0\text{ N/m}$$

(b) A 0.575 kg mass (weight 5.64 N) stretches the spring by

$$x = \frac{F}{k} = \frac{5.64\text{ N}}{49.0\text{ N/m}} = 1.15 \times 10^{-1}\text{ m} = 11.5\text{ cm}$$

2. From Newton's second law,

$$a = \frac{kx}{m} \quad \text{or} \quad 9.80\text{ m/s}^2 = k\frac{0.500\text{ m}}{3.00\text{ kg}}$$

From which, $k = 58.8$ N/m

3. $k = \dfrac{F}{x} = \dfrac{(4.00\text{ kg})(9.80\text{ m/s}^2)}{2.50 \times 10^{-2}\text{ m}} = 1570\text{ N/m}$

(a) $x = \dfrac{F}{k} = \dfrac{(1.50\text{ kg})(9.80\text{ m/s}^2)}{1570\text{ N/m}} = 9.38 \times 10^{-3}\text{ m} = 0.938\text{ cm}$

(b) $W = PE = \dfrac{1}{2}kx^2 = \dfrac{1}{2}(1570\text{ N/m})(4.00 \times 10^{-2}\text{ m})^2 = 1.25\text{ J}$

4. (a) The maximum elastic potential energy is stored when the spring is fully compressed. Thus,

$$PE(\text{max}) = \frac{1}{2}kx^2 = \frac{1}{2}(19.6\text{ N/m})(5.00 \times 10^{-2}\text{ m})^2 = 2.45 \times 10^{-2}\text{ J}$$

(b) We use $v = \sqrt{\dfrac{k}{m}(A^2 - x^2)}$ and note that the maximum velocity occurs when $x = 0$. Thus,

$$v\ (\text{max}) = \sqrt{\frac{19.6\text{ N/m}}{0.300\text{ kg}}}(5.00 \times 10^{-2}\text{ m}) = 4.04 \times 10^{-1}\text{ m/s} = 40.4\text{ cm/s}$$

5. We have: $PE_s)_{\text{at end}} + PE_g)_{\text{end}} = PE_g)_{\text{initial}}$. Taking $x = 0$ at the initial level of the trampoline, this becomes:

$$32\left(\frac{1}{2}kx^2\right) - mgx = mgh_i$$

or $32\left(\dfrac{1}{2}(5000\text{ N/m})x^2\right) - (40.0\text{ kg})(9.80\text{ m/s}^2)x = (40.0\text{ kg})(9.80\text{ m/s}^2)(1.93\text{ m})$

which reduces to: $(8.00 \times 10^4)x^2 - (392\text{ m})x - 756.6\text{ m}^2 = 0.$

5. (continued)
Using the quadratic formula (and taking the positive solution to get the lowest position of the person) gives:

$$x = 0.0997 \text{ m} = 9.97 \text{ cm}$$

6. From conservation of mechanical energy, we have: $\frac{1}{2}mv_f^2 = \frac{1}{2}kx_i^2$, or

$$\frac{1}{2}(0.100 \text{ kg})v_f^2 = \frac{1}{2}(200 \text{ N/m})(6.00 \times 10^{-2} \text{ m})^2$$

From which, $v_f = 2.68$ m/s.

7. $v = \frac{4.0 \text{ m}}{0.10 \text{ s}} = 40$ m/s and $\mu = \frac{0.20 \text{ kg}}{4.0 \text{ m}} = 5.0 \times 10^{-2}$ kg/m, so

$$F = \mu v^2 = (5.0 \times 10^{-2} \text{ kg/m})(40 \text{ m/s})^2 = 80 \text{ N}$$

8. The period of the motion is

$$T = 2\pi\sqrt{\frac{m}{k}} = 2\pi\sqrt{\frac{0.500 \text{ kg}}{30.0 \text{ N/m}}} = 0.258\pi \text{ s}$$

and the frequency is $f = \frac{1}{T} = 1.23$ Hz

We also know the amplitude of motion is 0.250 m. Thus, the equation of motion becomes

$$x = A\cos(2\pi ft) = (0.250 \text{ m})\cos 2.47\pi t$$

9. $T = 2\pi\sqrt{\frac{L}{g}}$

$$1.40 \text{ s} = 2\pi\sqrt{\frac{0.500 \text{ m}}{g}}$$ From which, $g = 10.1 \text{ m/s}^2$

10. $T = 2\pi\sqrt{\frac{L}{g}}$

$$1.00 \text{ s} = 2\pi\sqrt{\frac{L}{9.80 \text{ m/s}^2}}$$

From which, $L = 0.248$ m

11. $v = \lambda f = (0.0850 \text{ m})(2.00 \text{ Hz}) = 0.170$ m/s, and $t = \frac{d}{v} = \frac{10.0 \text{ m}}{0.170 \text{ m/s}} = 58.8$ s

12. $d = vt = (3.00 \times 10^8 \text{ m/s})(3.156 \times 10^7 \text{ s/y}) = 9.47 \times 10^{15}$ m

13. $v = \dfrac{4.00\text{ m}}{0.100\text{ s}} = 40.0\text{ m/s}$ and $\mu = \dfrac{0.200\text{ kg}}{4.00\text{ m}} = 5.00 \times 10^{-2}\text{ kg/m}$, so

    $$F = \mu v^2 = (5.00 \times 10^{-2}\text{ kg/m})(40.0\text{ m/s})^2 = 80.0\text{ N}$$

14. The hanging block is in equilibrium under the action of the tension in the string and its weight. From the first law, we find

    $$F = mg = (2.00\text{ kg})(9.80\text{ m/s}^2) = 19.6\text{ N}$$

    So, the mass per unit length of the string is

    $$\mu = \frac{F}{v^2} = \frac{19.6\text{ N}}{(15.0\text{ m/s})^2} = 8.71 \times 10^{-2}\text{ kg/m}$$

    Therefore, the mass of the string is

    $$\mu L = (8.71 \times 10^{-2}\text{ kg/m})(3.00\text{ m}) = 2.61 \times 10^{-1}\text{ kg}$$

15. From $v = \sqrt{\dfrac{F}{\mu}}$. The ratio of the wave speeds for the two cases (with μ constant) is

    $$\frac{v_2^{\,2}}{v_1^{\,2}} = \frac{F_2}{F_1}$$

    Therefore, if $F_2 = 2F_1$, we have

    $$v_2 = \sqrt{2}v_1 = (90.0\text{ m/s})\sqrt{2} = 127\text{ m/s}$$

16. (a) $E = \dfrac{1}{2}kA^2$, so if $A' = 2A$, $E' = \dfrac{1}{2}kA'^2 = \dfrac{1}{2}k(2A)^2 = 4E$ (E increases by factor of 4)

    (b) $v_{max} = \sqrt{\dfrac{k}{m}}A$, so if A is doubled, v_{max} is also doubled.

    (c) $a_{max} = \dfrac{k}{m}A$, so if A is doubled, a_{max} also doubles.

    (d) $T = 2\pi\sqrt{\dfrac{m}{k}}$ is independent of A, so the period is unchanged.

14 SOUND

PROBLEMS

1. What is the speed of sound in air at (a) 27.0°C, (b) 100°C, (c) 200°C?

2. What is the intensity level in decibels of a sound wave of intensity (a) 10^{-6} W/m^2 and (b) 1025 W/m^2?

3. What is the intensity of a sound whose intensity level is (a) 40 dB and (b) 100 dB?

4. The dB level at a distance of 2.0 m from a sound source is 100 dB. What is the dB level at distances of (a) 4.0 m, (b) 6.0 m, (c) 8.0 m?

5. A rock group is playing in a room. Sound emerging from the door spreads uniformly in all directions. If the intensity level of the music is 80.0 dB at a distance of 5.00 m from the door, at what distance is the music just barely audible to a person with a normal threshold of hearing? Disregard absorption.

6. At what speed should a supersonic aircraft fly so that the angle θ shown in Figure 14.1 will be (a) 40°, (b) 30°?

7. The sound level 3.0 m from a point source is 120 dB. At what distance will the sound level be 100 dB?

8. A string 50.0 cm long has a mass per unit length equal to 20.0×10^{-5} kg/m. To what tension should this string be stretched if its fundamental frequency is to be (a) 20.0 Hz and (b) 4500 Hz?

9. A stretched string is 160 cm long and has a linear density of 0.0150 g/cm. What tension in the string will result in a second harmonic of 460 Hz?

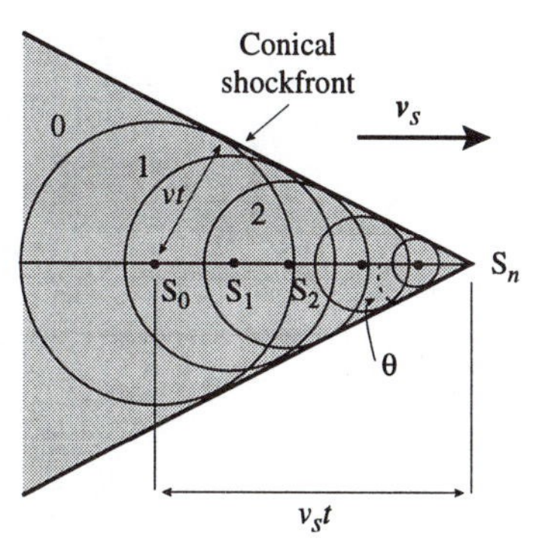

Figure 14.1

10. If the circumference of the mouth of a crystal goblet is 10.0 cm, at what fundamental frequency would an opera singer have to sing in order to set the glass into a resonant vibration?

11. A closed organ pipe is 3.00 m long. At what frequencies between 20.0 Hz and 20,000 Hz will this pipe resonate?

12. The range of a certain pipe organ is from 8.00 Hz to 30,000 Hz. What range of pipe lengths is necessary if they are (a) opens at both ends and (b) closed at one end? (Assume that the speed of sound is 343 m/s.)

13. A piano tuner strikes a 440-Hz tuning fork at the instant she strikes a piano key that should emit a tone of 440 Hz and hears a beat frequency of 2 Hz. What are the possible frequencies the piano key could be emitting?

14. What is the lowest frequency of the standing wave of sound that can be set up between two walls that are 8.00 m apart if the temperature is 22.0°C?

15. A standing wave is established in a string that is 240 cm long and fixed at both ends. The string vibrates in four segments when driven at 120 Hz. (a) Determine the wavelength. (b) What is the fundamental frequency?

16. A stretched string of length L is observed to vibrate in five equal segments when driven by a 630-Hz oscillator. What oscillator frequency will set up a standing wave so that the string vibrates in three segments?

CHAPTER 14 SOLUTIONS

1. We use $v = (331\text{ m/s})\sqrt{1 + \frac{T_C}{273}}$

   (a) For $T_C = 27.0°\text{C}$, $v = (331\text{ m/s})\sqrt{1 + \frac{27.0}{273}} = 347\text{ m/s}$

   (b) Using the same approach as in (a), we find $v = 387$ m/s at 100°C, and

   (c) $v = 436$ m/s at 200°C.

2. (a) $\beta = 10\log\left(\frac{I}{I_o}\right) = 10\log\left(\frac{10^{-6}}{10^{-12}}\right) = 10\log(10^6) = 60\text{ dB}$

   (b) $\beta = 10\log\left(\frac{I}{I_o}\right) = 10\log\left(\frac{10^{-5}}{10^{-12}}\right) = 10\log(10^7) = 70\text{ dB}$

3. (a) We use $\beta = 10\log\left(\frac{I}{I_o}\right)$

   $$40 = 10\log\left(\frac{I}{10^{-12}\text{ W/m}^2}\right)$$

   or, $4 = \log\left(\frac{I}{10^{-12}\text{ W/m}^2}\right)$ Taking the antilog of both sides gives

   $$\frac{I}{10^{-12}\text{ W/m}^2} = 10^4$$

   From which, $I = 10^{-8}\text{ W/m}^2$.

   (b) Following the approach used in (a), we find $I = 10^{-2}\text{ W/m}^2$.

4. At $r = 2.0$ m, $\beta = 100$ dB. Thus,

   $$100 = 10\log\left(\frac{I}{10^{-12}\text{ W/m}^2}\right)$$

   From which, we find $I = 10^{-2}\text{ W/m}^2$. The power emitted by the source is

   $$P = IA = I(4\pi r^2) = (10^{-2}\text{ W/m}^2)(4\pi(2.00\text{ m})^2) = 0.503\text{ W}$$

   The intensity is given by

   $$I = \frac{P}{A} = \frac{P}{4\pi r^2} = \frac{0.503\text{ W}}{4\pi r^2} \quad (1)$$

4. (continued)

(a) at $r = 4.0$ m, (1) gives $I = 2.5 \times 10^{-3}$ W/m^2. And $\beta = 10 \log\left(\frac{2.5 \times 10^{-3}}{10^{-12}}\right) = 94$ dB

(b) and (c) Using the approach outlined in (a), we find $\beta = 90.5$ dB at 6.0 m, and $\beta = 88$ dB at 8.0 m.

5. At $r = 5.00$ m from door, $I = I_5$, and at the desired distance $r = R$. Let $I = I_o$, the threshold of hearing. We have,

$$I = \frac{\text{const}}{r^2} \quad \text{or} \quad Ir^2 = \text{constant}$$

Therefore, $I_o R^2 = I_5 (5.00 \text{ m})^2$.

But, $80.0 = 10 \log \frac{I_5}{I_o}$, or

$$8.00 = \log \frac{I_5}{I_o} = \log \left(\frac{R}{5.00 \text{ m}}\right)^2 = 2 \log \frac{R}{5.00 \text{ m}}$$

which gives: $R = 5.00 \times 10^4$ m.

6. (a) $\sin \theta = \frac{v}{v_s}$

Thus, $\sin 40.0° = \frac{345 \text{ m/s}}{v_s}$

From which, $v_s = 537$ m/s

(b) $\sin 30.0° = \frac{345 \text{ m/s}}{v_s}$

From which, $v_s = 690$ m/s

7. If $\beta = 120$ dB at 3.0 m, then $I = 1.0$ W/m^2 at 3.0 m. Therefore,

$$P = IA = I(4\pi r^2) = (1.0 \text{ W/m}^2)(4\pi(3.0 \text{ m})^2) = 113 \text{ W}$$

At $\beta = 100$ dB, $I = 10^{-2}$ W/m^2. Thus,

$$r^2 = \frac{P}{4\pi I} = \frac{113 \text{ W}}{4\pi(10^{-2} \text{ W/m}^2)} = 900 \text{ m}^2$$

and $r = 30$ m

8. (a) For the fundamental mode of vibration, the length of the string

$L = \lambda/2 \quad \text{or} \quad \lambda = 2L = 100 \text{ cm}$

If $f = 20.0$ Hz, then $v = \lambda f = (1.00 \text{ m})(20.0 \text{ Hz}) = 20.0$ m/s

and $v = \sqrt{\dfrac{F}{\mu}}$

$$20.0 \text{ m/s} = \sqrt{\frac{F}{20.0 \times 10^{-5} \text{ kg/m}}}$$

From which, $F = 8.00 \times 10^{-2}$ N

(b) If $f = 4500$ Hz, then the wave speed is 4500 m/s, and using the approach of (a) we find,

$F = 4.05 \times 10^{3}$ N

9. In the second harmonic, $\lambda = L = 1.60$ m.

(Also, note that $\mu = 0.0150 \text{ g/cm} = 1.50 \times 10^{-3}$ kg/m.) Thus, if $f = 460$ Hz,

$v = \lambda f = (1.60 \text{ m})(460 \text{ m/s}) = 736 \text{ m/s}$

and $F = \mu v^2 = (1.50 \times 10^{-3} \text{ kg/m})(736 \text{ m/s})^2 = 813$ N

10. Resonance at the fundamental mode occurs when the circumference and the wavelength are the same. Thus,

$$f = \frac{v}{\lambda} = \frac{345 \text{ m/s}}{0.100 \text{ m}} = 3450 \text{ Hz}$$

11. The frequency of the fundamental mode is

$$f = \frac{v}{\lambda} = \frac{345 \text{ m/s}}{(4)(3.00 \text{ m})} = 28.8 \text{ Hz}$$

Thus, resonance occurs in the interval between 20.0 Hz and 20,000 Hz for frequencies given by $f_n = (28.8 \text{ Hz})n$ where n = any odd integer between 1 and 694.

12. (a) For an open pipe, $L = \dfrac{\lambda}{2}$ for the fundamental

For $f = 8.00$ Hz, $\lambda = \dfrac{v}{f} = \dfrac{343 \text{ m/s}}{8.00 \text{ Hz}} = 42.8$ m, and $L = 21.4$ m

For $f = 30000$ Hz, $\lambda = \dfrac{v}{f} = \dfrac{343 \text{ m/s}}{30000 \text{ Hz}} = 1.14 \times 10^{-2}$ m, and $L = 5.72 \times 10^{-3}$ m

(b) For the closed pipe, $L = \dfrac{3}{4}\lambda$ for the fundamental

12. (continued)

For $f = 8.00$ Hz, $L = \frac{3}{4}(42.8 \text{ m}) = 32.1 \text{ m}$

For $f = 30000$ Hz, $L = \frac{3}{4}(1.14 \times 10^{-2} \text{ m}) = 8.58 \times 10^{-3} \text{ m} = 8.58 \text{ mm}$

The range of lengths for the open pipe: 5.72 mm to 21.4 m and the range for the closed pipe: 8.58 mm to 32.1 m

13. The beat frequency is equal to the difference between the two frequencies beating together. Thus,

$$\Delta f = f_1 - f_2 = 2 \text{ Hz}$$

Thus, the two possible frequencies for the piano key are 442 Hz and 438 Hz.

14. At a temperature of 22.0°C, the speed of sound is 344 m/s. The lowest frequency corresponds to the longest wavelength. For the fundamental mode, λ and L are related as,

$$\lambda = 2L = 2(8.00 \text{ m}) = 16.0 \text{ m}$$

Thus, $f = \frac{v}{\lambda} = \frac{344 \text{ m/s}}{16.0 \text{ m}} = 21.5 \text{ Hz}$

15. (a) For a string vibrating in four segments, the wavelength of the wave is

$$\lambda = L/2 = 240 \text{ cm}/2 = 120 \text{ cm}$$

(b) The speed of the wave on the string is

$$v = \lambda f = (120 \text{ cm})(120 \text{ Hz}) = (120)^2 \text{ cm/s}$$

If the string vibrates in its fundamental mode, the wavelength of the wave is

$$\lambda = 2L = 2(240 \text{ cm}) = 480 \text{ cm}$$

Therefore, the frequency of the fundamental is

$$f = \frac{v}{\lambda} = \frac{(120)^2 \text{ cm/s}}{480 \text{ cm}} = 30 \text{ Hz}$$

16. If the string vibrates in five segments, the length of the string is $L = \frac{5}{2}\lambda$ or the wavelength is $\lambda = \frac{2}{5}L$. Similary, when the string vibrates in three segments, $\lambda' = \frac{2}{3}L$. Since the speed of transverse waves in the string is constant, $v = \lambda f = \lambda' f'$ giving

$$f' = \left(\frac{\lambda}{\lambda'}\right) f = \left(\frac{\frac{2}{5}L}{\frac{2}{3}L}\right) f = \left(\frac{3}{5}\right)(630 \text{ Hz}) = 378 \text{ Hz}$$

15 ELECTRIC FORCES AND ELECTRIC FIELDS

PROBLEMS

*Indicates intermediate level problems.

1. Two point charges of magnitude 3.0×10^{-9} C and 6.0×10^{-9} C are separated by a distance of 0.30 m. Find the electric force of repulsion between them.

2. Determine what the mass of a pair of protons would be if the gravitational and electrical forces between them were equal in magnitude.

3. Two point charges of equal magnitude repel each other with a force of 2.00 N when separated by 5.00 cm. Find the magnitude of the charge on each.

*4. The total charge on two spheres is 600 μC. The two spheres are placed 0.900 m apart and the force of repulsion between the two is 30.0 N. What is the charge on each sphere?

5. The electric force on a point charge of 5.0×10^{-9} C at some point is 3.8×10^{-3} N in the positive x direction. What is the magnitude of the electric field at this location?

6. The magnitude of the electric field at a certain location is 500 N/C and the field is directed east to west. (a) Find the magnitude and direction of the force acting on a proton placed at this point. (b) Repeat for an electron located at the point.

7. Find the magnitude and direction of the electric field at a distance of 10.0 cm from an electron.

8. The electric field at a distance of 0.800 m from a certain charge is found to have a magnitude of 200 N/C. What is the magnitude of the charge?

9. What is the strength of the electric field that would give a proton an acceleration equal to that of gravity?

*10. A point charge $q_1 = -3.50 \times 10^{-9}$ C is located on the y axis at $y = 0.120$ m, a charge $q_2 = -1.80 \times 10^{-9}$ C is located at the origin, and a charge $q_3 = 2.60 \times 10^{-9}$ C is located on the x axis at $x = -0.120$ m. Find the resultant force on q_3.

11. A proton is accelerated from rest by an electric field of 400 N/C. (a) Find its acceleration, (b) its speed after 10^{-8} s, and (c) its kinetic energy at this time.

*12. A proton is shot vertically upward with a speed of 2.00×10^5 m/s in a downward-directed electric field of 500 N/C. How high will it rise?

13. The Moon and Earth are bound together by gravity. If, instead, the force of attraction were the result of each having a charge of the same magnitude but opposite in sign, find the quantity of charge that would have to be placed on each to produce the required force.

14. An object with a net charge of 24 μC is placed in a uniform electric field of 610 N/C, directed vertically. What is the mass of this object if it "floats" in this electric field?

15. A constant electric field directed along the positive x axis has a strength of 2000 N/C. Find (a) the force exerted on the proton by the field, (b) the acceleration of the proton, and (c) the time required for the proton to reach a speed of 1.00×10^6 m/s, assuming it starts from rest.

CHAPTER 15 SOLUTIONS

1. $F = \frac{kq_1q_2}{r^2} = (8.99 \times 10^9 \text{ Nm}^2/\text{C}^2)\ \frac{(3.0 \times 10^{-9}\text{ C})(6.0 \times 10^{-9}\text{ C})}{(3.0 \times 10^{-1}\text{ m})^2} = 1.8 \times 10^{-6}\text{ N}$

2. We set $F_g = F_e$,

$$G\frac{m^2}{r^2} = k\frac{e^2}{r^2}\text{, which gives}$$

$$m = e\sqrt{\frac{k}{G}} = (1.60 \times 10^{-19}\ \text{C})\sqrt{\frac{8.99 \times 10^9\text{ Nm}^2/\text{C}^2}{6.67 \times 10^{-11}\text{ Nm}^2/\text{kg}^2}} = 1.86 \times 10^{-9}\text{ kg}$$

This is the mass of a single proton, and the mass for a pair would be just twice this $= 3.72 \times 10^{-9}$ kg.

3. Let us call Q the charge on each. The force of repulsion between them is $F = \frac{kQ^2}{r^2}$. From which,

$$Q = \sqrt{\frac{Fr^2}{k}} = \sqrt{\frac{(2.00\text{ N})(5.00 \times 10^{-2}\text{ m})^2}{(8.99 \times 10^9\text{ Nm}^2/\text{C}^2)}} = 7.45 \times 10^{-7}\text{ C}$$

*4. We are given that $Q_1 + Q_2 = 6.00 \times 10^{-4}$ C (1)

Also, we know $F = 30.0\text{ N} = \frac{kQ_1Q_2}{r^2}$

From which, we find

$$Q_1Q_2 = \frac{(30.0\text{ N})(0.900\text{ m})^2}{(8.99 \times 10^9\text{ Nm}^2/\text{C}^2)} = 2.70 \times 10^{-9}\text{ C}^2 \qquad (2)$$

We solve (1) and (2) simultaneously to find

$$Q_1 = 4.53 \times 10^{-6}\text{ C} \quad \text{and} \quad Q_2 = 5.95 \times 10^{-4}\text{ C}$$

5. $E = \frac{F}{q} = \frac{3.8 \times 10^{-3}\text{ N}}{5.0 \times 10^{-9}\text{ C}} = 7.6 \times 10^5$ N/C (in +x direction)

6. (a) $F = qE = (1.60 \times 10^{-19}\text{ C})(500\text{ N/C}) = 8.00 \times 10^{-17}$ N (force is in westward direction)

(b) The magnitude of the force is the same, but the direction is eastward for the electron.

7. $E = \frac{kq}{r^2} = \frac{(8.99 \times 10^9\text{ Nm}^2/\text{C}^2)(1.60 \times 10^{-19}\text{ C})}{(0.100\text{ m})^2} = 1.44 \times 10^{-7}$ N/C (toward electron)

8. $E = \dfrac{kq}{r^2}$

$200 \text{ N/C} = \dfrac{(8.99 \times 10^9 \text{ Nm}^2/\text{C}^2)q}{(0.800 \text{ m})^2}$

$q = 1.42 \times 10^{-8}$ C

9. The force needed would be

$$F = ma = (1.67 \times 10^{-27} \text{ kg})(9.80 \text{ m/s}^2) = 1.64 \times 10^{-26} \text{ N}$$

Thus, $E = \dfrac{F}{q} = \dfrac{1.64 \times 10^{-26} \text{ N}}{1.60 \times 10^{-19} \text{ C}} = 1.02 \times 10^{-7}$ N/C

10. The force exerted on q_3 by the charge q_1 on the y axis is directed toward q_1 (attractive). It makes an angle of 45.0° with the x axis. (See sketch.) Its magnitude is

$$F_1 = (8.99 \times 10^9 \text{ Nm}^2/\text{C}^2)\frac{(3.50 \times 10^{-9} \text{ C})(2.60 \times 10^{-9} \text{ C})}{(2.88 \times 10^{-2} \text{ m}^2)} = 2.84 \times 10^{-6} \text{ N}$$

The force exerted on q_3 by the -1.80×10^{-9} C charge at the origin is in the positive x direction (attractive), with magnitude,

$$F_2 = (8.99 \times 10^9 \text{ Nm}^2/\text{C}^2)\frac{(1.80 \times 10^{-9} \text{ C})(2.60 \times 10^{-9} \text{ C})}{(0.120 \text{ m})^2} = 2.93 \times 10^{-6} \text{ N}$$

The resultant force in the x direction is

$$F_x = F_{1x} + F_2 = 4.94 \times 10^{-6} \text{ N}$$

and the resultant force in the y direction is

$$F_y = F_{1y} = 2.01 \times 10^{-6} \text{ N}$$

The resultant is found from the Pythagorean theorem, as

$$F = \sqrt{(F_x)^2 + (F_y)^2} = 5.33 \times 10^{-6} \text{ N}$$

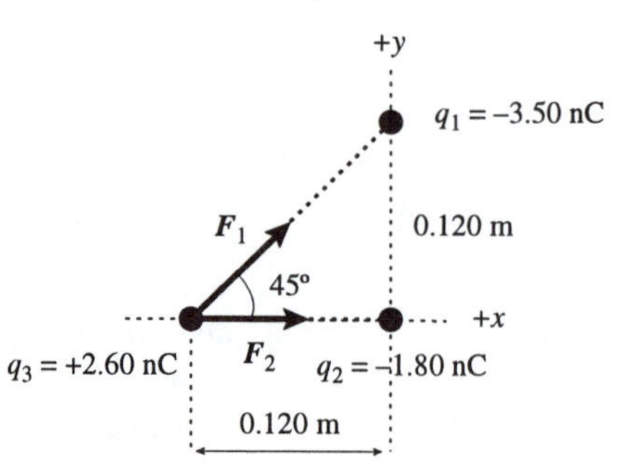

The angle θ is found from

$$\tan\theta = \frac{F_y}{F_x} \qquad \theta = 22.2°$$

11. (a) $F = qE = (1.60 \times 10^{-19} \text{ C})(400 \text{ N/C}) = 6.40 \times 10^{-17}$ N

$$a = \frac{F}{m} = \frac{6.40 \times 10^{-17} \text{ N}}{1.67 \times 10^{-27} \text{ kg}} = 3.83 \times 10^{10} \text{ m/s}^2$$

(b) $v = v_o + at$

$$v = 0 + (3.83 \times 10^{10} \text{ m/s}^2)(10^{-8} \text{ s}) = 383 \text{ m/s}$$

(c) $KE = \dfrac{1}{2}mv^2 = \dfrac{1}{2}(1.67 \times 10^{-27} \text{ kg})(383 \text{ m/s})^2 = 1.23 \times 10^{-22}$ J

12. The force on the proton is downward, and its magnitude is

$$F = qE = (1.60 \times 10^{-19}\ \text{C})(500\ \text{N/C}) = 8.00 \times 10^{-17}\ \text{N}$$

and the acceleration is also downward of magnitude

$$a = \frac{F}{m} = \frac{8.00 \times 10^{-17}\ \text{N}}{1.67 \times 10^{-27}\ \text{kg}} = 4.79 \times 10^{10}\ \text{m/s}^2$$

To find the maximum height, H, reached, we use

$$v_y^2 = v_{oy}^2 + 2ay$$

$$0 = (2.00 \times 10^5\ \text{m/s})^2 - 2(4.79 \times 10^{10}\ \text{m/s}^2)H$$

$$H = 4.18 \times 10^{-1}\ \text{m} = 41.8\ \text{cm}$$

13. If $F_e = F_g$, we have: $\frac{kQ^2}{r^2} = \frac{GMm}{r^2}$, where M is the mass of the earth, m the mass of the moon. Q is the charge that would have to be on each. We solve for Q to find: $Q = \sqrt{\frac{GMm}{k}}$, or

$$Q = \sqrt{\frac{(6.67 \times 10^{-11}\ \text{N m}^2/\text{kg}^2)(5.98 \times 10^{24}\ \text{kg})(7.36 \times 10^{22}\ \text{kg})}{(8.99 \times 10^9\ \text{N m}^2/\text{C}^2)}}$$

$$Q = 5.71 \times 10^{13}\ \text{C}$$

14. When the object "floats", the electrical and gravitational forces have the same magnitude. Thus, $mg = qE$ or $m = \frac{qE}{g}$, giving

$$m = \frac{(24 \times 10^{-6}\ \text{C})(610\ \text{N/C})}{9.80\ \text{m/s}^2} = 1.5 \times 10^{-3}\ \text{kg} = 1.5\ \text{g}$$

15. (a) The force on the proton is:

$$F = qE = (1.60 \times 10^{-19})(2.00 \times 10^3\ \text{N/C}) = 3.20 \times 10^{-16}\ \text{N} \quad (+x\ \text{direction})$$

(b) The acceleration is found from Newton's second law:

$$a = \frac{F}{m} = \frac{3.20 \times 10^{-16}\ \text{N}}{1.67 \times 10^{-27}\ \text{kg}} = 1.91 \times 10^{11}\ \text{m/s}^2 \quad (+x\ \text{direction})$$

(c) $v = v_o + at$ gives: $10^6\ \text{m/s} = 0 + (1.91 \times 10^{11}\ \text{m/s}^2)t$, and

$$t = 5.23 \times 10^{-6}\ \text{s}$$

16 ELECTRICAL ENERGY AND CAPACITANCE

PROBLEMS

*Indicates intermediate level problems.

1. Show that the units of N/C and V/m are equivalent.

2. A capacitor consists of two parallel plates separated by a distance of 0.300 mm. If a 20.0-V potential difference is maintained between those plates, calculate the electric field strength in the region between the plates.

3. A proton is between two plates, separated by a distance of 2.0 cm, across which there is a potential difference of 5000 V. (a) Find the electrical force exerted on the proton, and (b) its acceleration.

4. What potential difference is needed to stop an electron with an initial speed of 4.20×10^5 m/s?

5. At what distance from a point charge of 6.0 μC would the potential equal 2.7×10^4 V?

6. In the Bohr model of the hydrogen atom an electron circles a proton in an orbit of radius 0.510×10^{-10} m. Find the potential at the position of the electron.

7. To what difference in potential would a 2.0-μF capacitor have to be connected in order to store 98 μC of charge?

8. If the area of the plates of a parallel plate capacitor is doubled while the spacing between the plates is halved, how is the capacitance affected?

9. (a) Determine the equivalent capacitance for the capacitor network shown in Figure 16.1. (b) If the network is connected to a 12-V battery, calculate the potential difference across each capacitor and the charge on each capacitor.

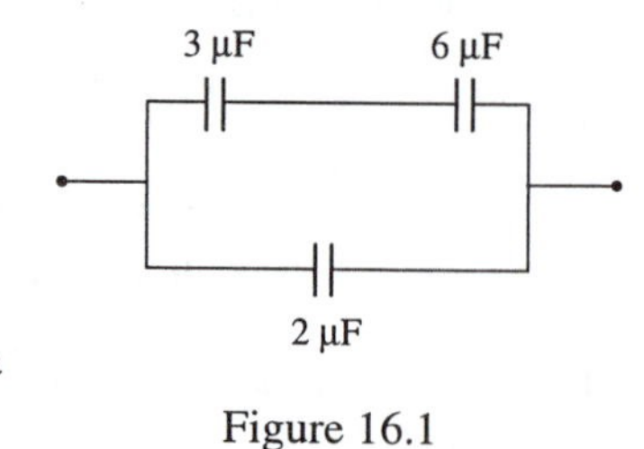

Figure 16.1

10. (a) A 3.00-μF capacitor is connected to a 12.0-V battery. How much energy is stored in the capacitor? (b) If the capacitor had been connected to a 6.00-V battery how much energy would have been stored?

11. A capacitor is connected to a 120-V source and holds a charge of 36.0 μC. (a) What is the capacitance of the capacitor? (b) Find the energy stored by the capacitor by the use of at least two different equations.

12. Find the capacitance of a parallel-plate capacitor that uses Bakelite as a dielectric if the plates of each have an area of 5.00 cm^2 and the plate separation is 2.00 mm.

13. An electron in the beam of a typical television picture tube is accelerated through a potential difference of 20 000 V before it strikes the face of the tube. What is the energy of this electron, in electron volts, and what is its speed when it strikes the screen?

14. Find the equivalent capacitance of the group of capacitors shown in Figure 16.2.

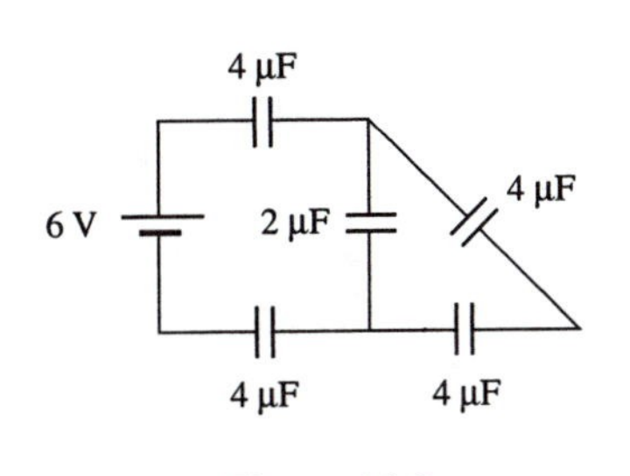

Figure 16.2

15. Find the total energy stored by the group of capacitors in Figure 16.2.

CHAPTER 16 SOLUTIONS

1. $1\frac{\text{volt}}{\text{meter}} = \frac{1\text{ joule/coul}}{1\text{ meter}} = \frac{1\text{ joule}}{1\text{ coul meter}} = \frac{1\text{ N meter}}{1\text{ coul meter}} = 1\ \frac{\text{N}}{\text{C}}$

2. $\Delta V = -Ed = -E(3.00 \times 10^{-4}\text{ m}) = -20.0\text{ V}$

   so $E = 6.67 \times 10^4\text{ V/m}$

3. (a) The strength of the electric field is

   $$E = \frac{\Delta V}{d} = \frac{5.0 \times 10^3\text{ V}}{2.0 \times 10^{-2}\text{ m}} = 2.5 \times 10^5\text{ N/C}$$

   and $F = qE = (1.60 \times 10^{-19}\text{ C})(2.5 \times 10^5\text{ N/C}) = 4.0 \times 10^{-14}\text{ N}$

   (b) $a = \frac{F}{m} = \frac{4.0 \times 10^{-14}\text{ N}}{1.67 \times 10^{-27}\text{ kg}} = 2.4 \times 10^{13}\text{ m/s}^2$

4. $W = \Delta KE = -q\Delta V$

   $$0 - \frac{1}{2}(9.11 \times 10^{-31}\text{ kg})(4.20 \times 10^5\text{ m/s})^2 = (-1.60 \times 10^{-19}\text{ C})\Delta V$$

   From which, $\Delta V = -0.502\text{ V}$

5. $V = k\frac{q}{r}$, or $r = \frac{kq}{V} = \frac{(8.99 \times 10^9\text{ N m}^2/\text{C}^2)(6.0 \times 10^{-6}\text{ C})}{2.7 \times 10^4\text{ N m/C}} = 2.0\text{ m}$

6. $V = k\frac{q}{r} = \frac{(8.99 \times 10^9\text{ N m}^2/\text{C}^2)(1.60 \times 10^{-19}\text{ C})}{0.510 \times 10^{-10}\text{ m}} = 28.2\text{ V}$

7. $\Delta V = \frac{Q}{C} = \frac{98\ \mu\text{C}}{2.0\ \mu\text{F}} = 49\text{ V}$

8. $C = \frac{\varepsilon_0 A}{d}$, so if $A_2 = 2A_1$ and $d_2 = \frac{1}{2}d_1$, then

   $$C_2 = \frac{\varepsilon_0 A_2}{d_2} = \frac{\varepsilon_0 2A_1}{\frac{1}{2}d_1} = 4\ \frac{\varepsilon_0 A_1}{d_1}$$

   or $C_2 = 4C_1$ (The capacitance is quadrupled.)

9. The circuit reduces as shown below.

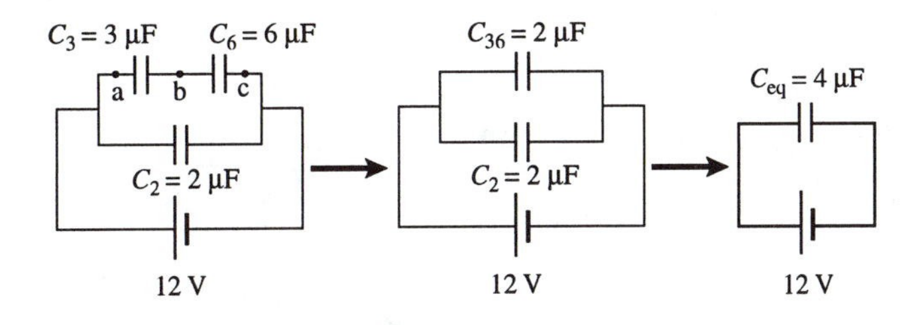

(b) $Q_2 = C_2 \Delta V_2 = (2\ \mu F)(12\ V) = 24\ \mu C$,

$Q_{36} = C_{36} \Delta V_2 = (2\ \mu F)(12\ V) = 24\ \mu C = Q_3 = Q_6$

Thus, $\Delta V_3 = \dfrac{Q_3}{C_3} = \dfrac{24\ \mu C}{3\ \mu F} = 8\ V$,

$$\Delta V_6 = \frac{Q_6}{C_6} = \frac{24\ \mu C}{6\ \mu F} = 4\ V$$

10. (a) $W = \dfrac{1}{2} C \Delta V^2 = \dfrac{1}{2}(3.00\ \mu F)(12.0\ V)^2 = 2.16 \times 10^{-4}\ J$

(b) $W = \dfrac{1}{2} C \Delta V^2 = \dfrac{1}{2}(3.00\ \mu F)(6.00\ V)^2 = 5.40 \times 10^{-5}\ J$

11. (a) $C = \dfrac{Q}{\Delta V} = \dfrac{36.0\ \mu C}{120\ V} = 0.300\ \mu F$

(b) $W = \dfrac{1}{2} Q \Delta V = \dfrac{1}{2}(36.0\ \mu C)(120\ V) = 2.16 \times 10^{-3}\ J$

or, $W = \dfrac{1}{2} C \Delta V^2 = \dfrac{1}{2}(0.300\ \mu F)(120\ V)^2 = 2.16 \times 10^{-3}\ J$

or, $W = \dfrac{1}{2}\dfrac{Q^2}{C} = \dfrac{1}{2}\dfrac{(36.0\ \mu C)^2}{(0.300\ \mu F)} = 2.16 \times 10^{-3}\ J$

12. $C = \dfrac{\kappa \varepsilon_0 A}{d} = \dfrac{4.9(8.85 \times 10^{-12}\ F/m)(5.00 \times 10^{-4}\ m^2)}{2.00 \times 10^{-3}\ m} = 1.08 \times 10^{-11}\ F = 10.8\ pF$

13. From conservation of energy, $K_f + qV_f = K_i + qV_i$, or $K_f = q(V_i - V_f)$.

Thus, $K_f = (1\ e)(20{,}000\ V) = 20{,}000\ eV$. Then $K_f = \dfrac{1}{2} m v_f^2 = 20{,}000\ eV$, so

$$v_f^2 = \frac{2(20000\ eV)(1.60 \times 10^{-19}\ J/eV)}{9.10 \times 10^{-31}\ kg} = 7.03 \times 10^{15}\ m^2/s^2$$

and $v_f = 8.39 \times 10^7\ m/s$.

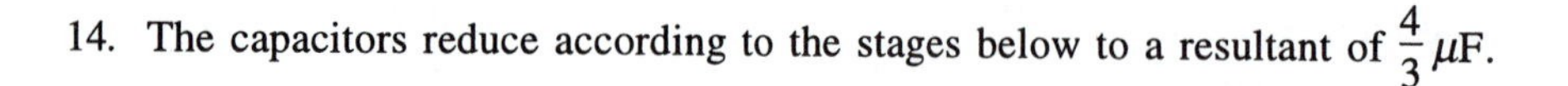
14. The capacitors reduce according to the stages below to a resultant of $\frac{4}{3}\,\mu$F.

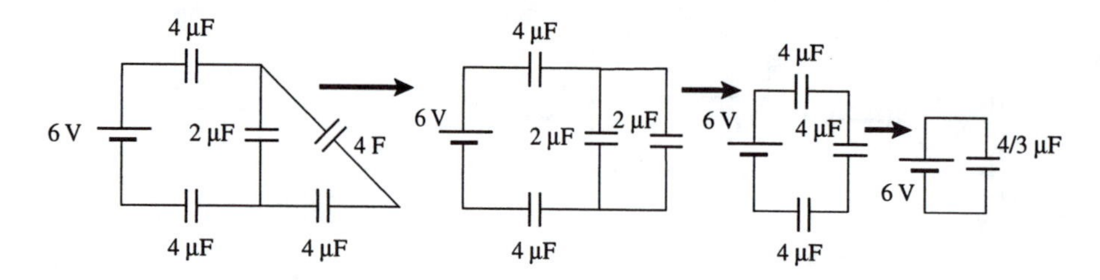

15. The charge stored on the equivalent capacitance is

$$Q = C_{eq}\Delta V = \left(\frac{4}{3} \times 10^{-6}\ \text{F}\right)(6.0\ \text{V}) = 8.0 \times 10^{-6}\ \text{C}$$

Thus, $W_{total} = \frac{1}{2} Q\Delta V = \frac{1}{2}(8.0 \times 10^{-6}\ \text{C})(6.0\ \text{V}) = 2.4 \times 10^{-5}\ \text{J}$

17 CURRENT AND RESISTANCE

PROBLEMS

*Indicates intermediate level problems.

1. If Avogadro's number of electrons pass by a given cross-sectional area in one hour, find the current in the conductor.

2. A metal wire 1.00 mm in diameter contains 2.50×10^{22} free electrons per cubic centimeter. If the electrons travel through the wire with an average drift speed of 0.500 mm/s, what is the current in the wire?

3. Find the resistance of a piece of silver 1.00 m long if its cross-sectional area is 1.00×10^{-5} m^2.

4. Calculate the diameter of a 2.00-cm length of tungsten filament in a small lightbulb if its resistance is 0.050 Ω.

5. A lightbulb has a resistance of 240 Ω when operating at voltage of 120 V. What is the current through the lightbulb?

6. A sales representative promoting energy-efficient products advertises that a certain product has a resistance of 20 Ω and carries a current of only 0.6 A when connected to a 120-V source. Would you buy this product? Why or why not?

7. Calculate the percentage change in the resistance of a carbon filament when it is heated from 20°C to 160°C.

8. How much would the temperature of a copper wire have to be increased to raise its resistance by 20% over the value it had at 20.0°C?

9. At what temperature will tungsten have a resistivity four times the resistivity of copper at room temperature?

10. A toaster is rated at 600 W when connected to a 120-V source. What current does the toaster carry and what is its resistance?

11. How much current is being supplied by a 240-V generator delivering 120 kW of power?

12. The sticker on a compact disk player says that it draws 300 mA of current at 9.00 V. What power does it dissipate?

13. A 0.250-hp motor is connected to a 120-V line. What current is carried by the motor?

14. If electricity costs 8 cents per kWh, estimate how much it costs a person to dry his or her hair with a 1500-W blow dryer during a year's time.

15. If an electric stove is capable of heating 500 cm^3 of water from 20.0°C to boiling in 10.0 minutes, what is the power rating of the stove? Assume 50% efficiency.

16. A typical aluminum wire has about 1.5×10^{20} free electrons in every centimeter of its length, and the drift speed of these electrons is about 0.030 cm/s. (a) How many electrons pass through a cross-sectional area of the wire every second? (b) What is the current in the wire?

17. A 2.0-m piece of iron wire carries a current of 0.25 A when connected to a 6.0-V battery. What length of gold wire, with the same radius, would carry the same current when connected to the battery?

CHAPTER 17 SOLUTIONS

1. $\Delta Q = N_a e = (6.02 \times 10^{23}$ electrons$)(1.60 \times 10^{-19}$ C/electron$) = 9.63 \times 10^4$ C

   From which, $I = \frac{\Delta Q}{\Delta t} = \frac{9.63 \times 10^4 \text{ C}}{3600 \text{ s}} = 26.8$ A

2. $I = nqAv_d = (2.50 \times 10^{28}\ 1/\text{m}^3)(1.60 \times 10^{-19}\ \text{C})(5.00 \times 10^{-4}\ \text{m/s})(7.85 \times 10^{-7}\ \text{m}^2) = 1.57$ A

3. $R = \frac{\rho L}{A} = \frac{(1.59 \times 10^{-8}\ \Omega\ \text{m})(1.00\ \text{m})}{1.00 \times 10^{-5}\ \text{m}^2} = 1.59 \times 10^{-3}\ \Omega$

4. $A = \frac{\rho L}{R} = \frac{(5.6 \times 10^{-8}\ \Omega\ \text{m})(2.00 \times 10^{-2}\ \text{m})}{5.00 \times 10^{-2}\ \Omega} = 2.24 \times 10^{-8}\ \text{m}^2$

   But $A = \pi d^2/4$. From which, $d = 1.69 \times 10^{-4}$ m

5. $I = \frac{\Delta V}{R} = \frac{120\ \text{V}}{240\ \Omega} = 0.500$ A

6. If the voltage is 120 V and $R = 20\ \Omega$, then

   $I = \frac{\Delta V}{R} = \frac{120\ \text{V}}{20\ \Omega} = 6.0$ A

   (His claim violates Ohm's law.)

7. % change $= \frac{R - R_o}{R_o} 100\% = \left(\frac{R}{R_o} - 1\right)100$, and $\frac{R}{R_o} = 1 + \alpha(\Delta T)$

   Therefore, % change $= \alpha(\Delta T)100 = 100\ (-0.5 \times 10^{-3}\ °\text{C}^{-1})(140°\text{C}) = -7\%$

8. We are given $R = 1.2R_o$. Thus,

   $R = R_o(1 + \alpha(\Delta T))$ becomes

   $1.2R_o = R_o(1 + \alpha(\Delta T))$

   or, $\alpha \Delta T = 0.200$

   Thus, $\Delta T = \frac{2.00 \times 10^{-1}}{3.90 \times 10^{-3}\ °\text{C}^{-1}} = 51.3°\text{C}$

   Therefore, the final temperature is 71.3°C.

9. We use, $\rho = \rho_o(1 + \alpha(\Delta T))$

   For tungsten, $\rho_o = 5.60 \times 10^{-8}\ \Omega$ m and $\alpha = 4.5 \times 10^{-3}\ °C^{-1}$

   If $\rho = 4(\rho_o)_{copper} = 4(1.7 \times 10^{-8}\ \Omega\ \text{m}) = 6.8 \times 10^{-8}\ \Omega$ m, then

   $$6.8 \times 10^{-8}\ \Omega\ \text{m} = 5.6 \times 10^{-8}\ \Omega\ \text{m}(1 + 4.5 \times 10^{-3}\ °C^{-1}\Delta T)$$

   We find, $\Delta T = 47.6°C$

   and $T = 67.6°C$

10. $I = \frac{P}{\Delta V} = \frac{600\ \text{W}}{120\ \text{V}} = 5.0\ \text{A}$

    and, $R = \frac{\Delta V}{I} = \frac{120\ \text{V}}{5.0\ \text{A}} = 24\ \Omega.$

11. $I = \frac{P}{\Delta V} = \frac{120 \times 10^3\ \text{W}}{240\ \text{V}} = 500\ \text{A}$

12. $P = \Delta VI = (9.00)(0.300\ \text{A}) = 2.70\ \text{W}$

13. $P = 0.250$ hp $= 186.5$ W. Thus, $I = \frac{P}{\Delta V} = \frac{186.5\ \text{W}}{120\ \text{V}} = 1.55\ \text{A}$

14. We shall assume five minutes of use per day. The energy required is

    $$W = Pt = (1.5\ \text{kW})\left(\frac{5\ \text{min}}{\text{day}}\right)\left(\frac{1\ \text{h}}{60\ \text{min}}\right)\left(\frac{365\ \text{days}}{\text{yr}}\right) = 45.7\ \text{kWh (per year)}$$

    Thus, cost = (8 cents/kWh)(45.7 kWh/yr) = 365 cents per year = $3.65 /year

15. 500 cm^3 of water = 500 g of water = 0.500 kg

    energy output = $mc\Delta T = (500\ \text{g})(1\ \text{cal/g °C})(80.0°C) = 4.00 \times 10^4\ \text{cal} = 1.674 \times 10^5\ \text{J}$

    energy out = (eff)(energy in)

    so, energy in $= \frac{\text{energy out}}{\text{eff}} = \frac{1.674 \times 10^5\ \text{J}}{0.500} = 3.35 \times 10^5\ \text{J}$

    and $P = \frac{\text{energy input}}{\text{time}} = \frac{3.35 \times 10^5\ \text{J}}{600\ \text{s}} = 558\ \text{W}$

16. (a) In one second, all the electrons in a 0.030 cm length of the wire must pass the cross-section. Therefore,

$$n = (1.5 \times 10^{20} \text{ electrons/cm})(0.030 \text{ cm/s}) = 4.5 \times 10^{18} \text{ electrons/s}$$

(b) $I = \dfrac{\Delta Q}{\Delta t} = ne = (4.5 \times 10^{18} \text{ electrons/s})(1.60 \times 10^{-19} \text{ C/electron})$, or

$$I = 0.72 \text{ C/s} = 0.72 \text{ A}.$$

17. The gold wire must have the same resistance as the iron wire. Thus,

$$\frac{\rho_{gold} L_{gold}}{A_{gold}} = \frac{\rho_{iron} L_{iron}}{A_{iron}}$$

But, $A_{gold} = A_{iron}$.

Therefore, $L_{gold} = \left(\dfrac{\rho_{iron}}{\rho_{gold}}\right) L_{iron} = \left(\dfrac{10 \times 10^{-8}\ \Omega\ \text{m}}{2.44 \times 10^{-8} \Omega\ \text{m}}\right)(2.0 \text{ m}) = 8.2 \text{ m}.$

18 DIRECT CURRENT CIRCUITS

PROBLEMS

1. A 9.0-Ω and a 6.0-Ω resistor are connected in series to a power supply. The current in the circuit is found to be 1/3 A. What is the voltage setting of the power supply?

2. A television repairman needs a 100-Ω resistor to repair a malfunctioning set. He is temporarily out of a resistor of this value. All he has in his tool box is a 500-Ω resistor and two 250-Ω resistors. How can the desired resistance be obtained from the resistors on hand?

3. A 9.00-Ω and a 6.00-Ω resistor are connected in parallel across a power supply. If the current through the equivalent resistor is 1/3 A, find the voltage setting of the power supply.

4. Three resistors (10.0 Ω, 20.0 Ω, 30.0 Ω) are connected in parallel. The total current through this network is 5.00 A. (a) What is the voltage drop across the network? (b) What is the current in each resistor?

5. A length of wire is cut into five equal pieces. The five pieces are then connected in parallel, with the resulting resistance being 2.00 Ω. What was the resistance of the original length of wire?

6. A pair of 1.50-V penlight batteries in series power a transistor radio. The batteries hold a total charge of 180 C. How long will they last if the effective resistance of the radio is 300 Ω?

7. Determine the current in the circuit of Figure 18.1 and the voltage drop across the 400-Ω resistor.

8. How many 100-W lightbulbs can you connect in a 120-V house circuit without tripping a 20.0-A circuit breaker?

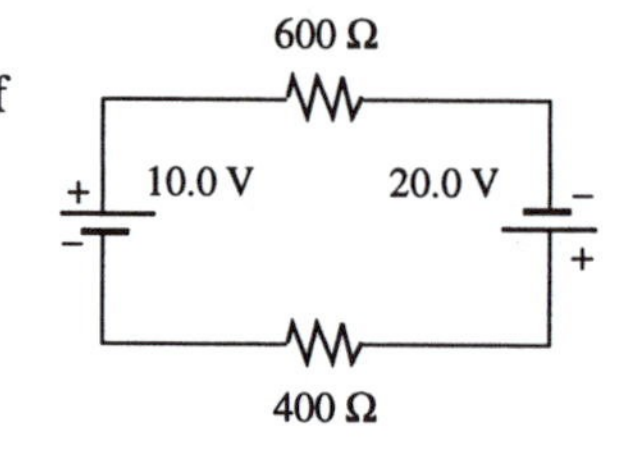

Figure 18.1

9. An 8-foot extension cord has two 18-gauge copper wires, each having a diameter of 1.024 mm. (a) How much power does this cord dissipate when carrying a current of 1.00 A? (b) How much power does this cord dissipate when carrying a current of 10.0 A?

10. If you have a 2-Ω, a 4-Ω, and a 6-Ω resistor, how many different resistance values can you obtain from the set? Show how you would connect them to get each resistance value.

11. In a certain circuit, a current of 1.00 A is drawn from a battery. The current then divides and passes through two resistors in parallel. One of the resistors has a value of 41.0 Ω, and the current through it is 0.100 A. What is the value of the other resistor?

12. When a battery of unknown emf is connected to a 5.0-Ω resistor, the current in the circuit is 0.30 A. If the battery is now connected to an 8.0-Ω resistor, the current is 0.20 A. What are the emf of the battery and its internal resistance?

CHAPTER 18 SOLUTIONS

1. The equivalent resistance is,

$$R_{eq} = R_1 + R_2 = 9.0\ \Omega + 6.0\ \Omega = 15\ \Omega$$

Therefore, the voltage setting of the power supply is

$$\Delta V = IR_{eq} = \left(\frac{1}{3}\text{ A}\right)(15\ \Omega) = 5.0\text{ V}$$

2. Consider them all in parallel.

$$\frac{1}{R_{eq}} = \frac{1}{R_1} + \frac{1}{R_2} + \frac{1}{R_3} = \frac{1}{500\ \Omega} + \frac{1}{250\ \Omega} + \frac{1}{250\ \Omega}$$

From which, $R_{eq} = 100\ \Omega$.

3. The equivalent resistance of the two resistors in parallel is

$$\frac{1}{R_{eq}} = \frac{1}{R_1} + \frac{1}{R_2} = \frac{1}{9.00\ \Omega} + \frac{1}{6.00\ \Omega}$$

Which yields, $R_{eq} = \frac{18}{5}\ \Omega$.

Therefore, $\Delta V = IR_{eq} = \left(\frac{1}{3}\text{ A}\right)\left(\frac{18\ \Omega}{5}\right) = \frac{6}{5}\text{ V} = 1.20\text{ V}$

4. $$\frac{1}{R_{eq}} = \frac{1}{R_1} + \frac{1}{R_2} + \frac{1}{R_3} = \frac{1}{10.0\ \Omega} + \frac{1}{20.0\ \Omega} + \frac{1}{30.0\ \Omega}$$

From which, $R_{eq} = \frac{60}{11}\ \Omega$.

(a) The voltage drop across the parallel branch is

$$\Delta V = IR_{eq} = (5.00\text{ A})\left(\frac{60}{11}\ \Omega\right) = 27.3\text{ V}$$

(b) $I_{10} = \frac{\Delta V}{10.0\ \Omega} = \frac{27.3\text{ V}}{10.0\ \Omega} = 2.73\text{ A}$

$$I_{20} = \frac{\Delta V}{20.0\ \Omega} = \frac{27.3\text{ V}}{20.0\ \Omega} = 1.36\text{ A}$$

$$I_{30} = \frac{\Delta V}{30.0\ \Omega} = \frac{27.3\text{ V}}{30.0\ \Omega} = 0.910\text{ A}$$

5. Let R be the resistance of the original length. When cut, each piece has resistance $R/5$. When these are placed in parallel,

$$\frac{1}{R_{eq}} = \frac{1}{\frac{R}{5}} + \frac{1}{\frac{R}{5}} + \frac{1}{\frac{R}{5}} + \frac{1}{\frac{R}{5}} + \frac{1}{\frac{R}{5}} = \frac{5}{\frac{R}{5}} = \frac{25}{R}, \text{ so } R_{eq} = \frac{R}{25} = 2\ \Omega$$

Thus, $R = 50\ \Omega$.

6. The current in the circuit is

$$I = \frac{3.00\text{ V}}{300\ \Omega} = 0.0100\text{ A} = 0.0100\text{ C/s}$$

Thus, if the available charge = 180 C, the batteries will last

$$t = \frac{180\text{ C}}{0.0100\text{ C/s}} = 18{,}000\text{ s} = 5\text{ h}$$

7. The loop rule becomes

$$10.0\text{ V} - (600\ \Omega)I + 20.0\text{ V} - (400\ \Omega)I = 0$$

and $I = 0.030\text{ A} = 30\text{ mA}$

also $\Delta V = IR = (0.030\text{ A})(400\ \Omega) = 12.0\text{ V}$

8. The current drawn by a single bulb is

$$I = \frac{P}{V} = \frac{100\text{ W}}{120\text{ V}} = 0.833\text{ A}$$

Thus, the number of bulbs that a 20.0 A circuit can supply is

$$n = \frac{20.0\text{ A}}{0.833\text{ A}} = 24\text{ bulbs}$$

9. Total length of wire = 16.0 ft = 4.88 m, and the cross-sectional area $A = 8.24 \times 10^{-7}\text{ m}^2$

Thus, $R = \frac{\rho L}{A} = \frac{(1.7 \times 10^{-8}\ \Omega\text{ m})(4.88\text{ m})}{8.24 \times 10^{-7}\text{ m}^2} = 0.101\ \Omega$

(a) $P = I^2R = (1.00\text{ A})(0.101\ \Omega) = 0.101\text{ W}$

(b) $P = I^2R = (10.0\text{ A})(0.101\ \Omega) = 10.1\text{ W}$

10. Using one resistor, the following values are possible.

    $2\ \Omega$, $4\ \Omega$, and $6\ \Omega$ (3 distinct values)

    Using two resistors, either in series or in parallel, yields the following possible values.

    $6\ \Omega$, $8\ \Omega$, $10\ \Omega$, $1.33\ \Omega$, $1.5\ \Omega$, and $2.4\ \Omega$ (5 distinct values, and one repeat)

    Using all three resistors in all the possible combinations, yields

    $12\ \Omega$, $4.4\ \Omega$, $5.5\ \Omega$, $7.33\ \Omega$, $1.09\ \Omega$ (5 distinct values)

    Total number of distinct values = 13

11. Apply the junction rule at point a.

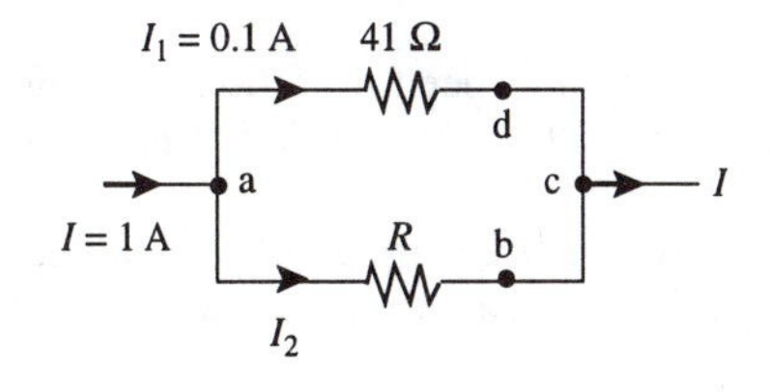

    $I = I_1 + I_2$

    or $1.00\text{ A} = 0.100\text{ A} + I_2$

    gives, $I_2 = 0.900\text{ A}$

    Now use the loop rule around loop abcda.

    $-RI_2 + (41.0\ \Omega)I_1 = 0$

    Knowing I_1 and I_2 we find, $R = 4.56\ \Omega$

12. Using the loop rule in case 1, we have:

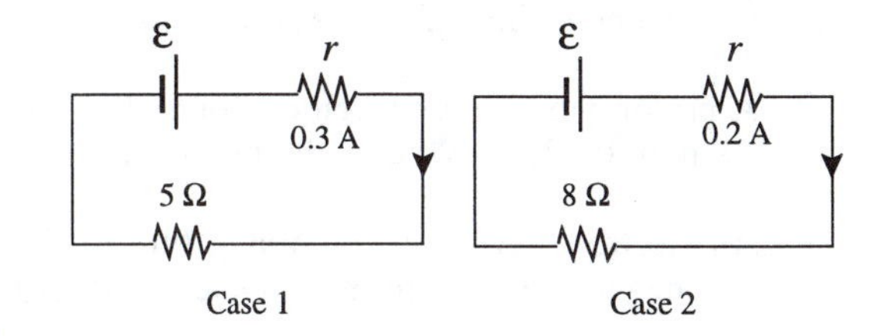

    $\mathcal{E} - (0.30\text{ A})r - (0.30\text{ A})(5.0\ \Omega) = 0$ (1)

    Once again, use the loop rule for case 2:

    $\mathcal{E} - (0.20\text{ A})r - (0.20\text{ A})(8.0\ \Omega) = 0$ (2)

    Equations (2) and (1) are solved together to find:

    $r = 1.0\ \Omega$ and $\mathcal{E} = 1.8\text{ V}$

19 MAGNETISM

PROBLEMS

*Indicates intermediate level problems.

1. A proton is moving at right angles to a magnetic field of 2.00 T. What speed does the proton have if the magnetic force on it has a magnitude of 6.00×10^{-11} N?

2. Consider the electron near the magnetic equator. In which direction will it tend to be deflected if its velocity is directed (a) downward, (b) northward, (c) westward, or (d) southeastward?

3. A proton moving with a speed of 5.00×10^7 m/s through a magnetic field of 2.00 T experiences a magnetic force of 3.00×10^{-12} N. What is the angle between the proton's velocity and the field?

4. Calculate the force on a 2.0-m length of conductor carrying a current of 10 A in a region where a uniform magnetic field has a magnitude of 1.2 T and is directed perpendicular to the conductor.

5. A vertical wire carries a current of 10 A directed upward at a location where the magnetic field of the Earth is horizontal and has a value of 5.0×10^{-5} T. If the wire is 20 m long, find the magnitude and direction of the net magnetic force on it.

6. A rectangular coil of 500 turns has dimensions of 4.00 cm by 5.00 cm and is suspended in a magnetic field of 0.650 T. What is the current in the coil if the maximum torque exerted on it by the magnetic field is 0.180 N · m?

7. A proton moves with a velocity of 2.00×10^5 m/s perpendicular to a uniform magnetic field of strength 0.200 T. What is the radius of the path?

8. An electron travels in a circular orbit of radius 1.77 m in a region of uniform magnetic field. If the magnetic field strength doubles, what is the new radius of the orbit?

9. What is the radius of the circular path of an electron with a kinetic energy of 1.60×10^{-19} J moving perpendicular to a 0.200-T magnetic field?

10. A singly charged positive ion moving with a speed of 4.60×10^5 m/s leaves a spiral track of radius 7.94 mm in a photograph along a direction perpendicular to the magnetic field of a bubble chamber. The magnetic field applied for the photograph has a magnitude of 1.80 T. Compute the mass of this particle.

11. Find the magnetic field strength at a distance of (a) 10 cm from a long, straight wire carrying a current of 5 A. Repeat the calculation for distances of (b) 50 cm and (c) 2 m.

12. Two parallel conductors are each 0.5 m long and carry 10-A currents in opposite directions. (a) What center-to-center separation must the conductors have if they are to repel each other with a force of 1.0 N? (b) Is this physically possible?

13. Two long, straight wires separated by a distance of 0.300 m carry currents in the same direction. If the current in one wire is 5.00 A and the current in the other is 8.00 A, find the magnitude and direction of the force that one exerts on a 2.00-m length of the other.

14. What current in a solenoid 15.0 cm long wound with 100 turns would produce a magnetic field at the center equal to that of the Earth, which is about 5.00×10^{-5} T?

15. Consider the solenoid in Problem 14. If it carries the same current, how many turns must it have to increase the strength of the magnetic field by a factor of 3?

16. A singly charged positive ion with a mass of 6.68×10^{-27} kg moves clockwise in a circular path of radius 3.00 cm with a speed of 1.00×10^{4} m/s. Find the direction and strength of the magnetic field.

CHAPTER 19 SOLUTIONS

1. From, $F = qvB \sin \theta$

   we find, $v = \dfrac{F}{qB} = \dfrac{6.00 \times 10^{-11}\ \text{N}}{(1.60 \times 10^{-19}\ \text{C})(2.00\ \text{T})} = 1.88 \times 10^{8}\ \text{m/s}$

2. Use the right hand rule, realizing that, at the magnetic equator, B is parallel to the surface of Earth and directed northward and that the electron has a negative charge to find the following answers:

   (a) westward
   (b) no deflection, v is parallel to B
   (c) upward
   (d) downward

3. From $F = qvB \sin \theta$, we have

   $$\sin \theta = \frac{F}{qvB} = \frac{3.00 \times 10^{-12}\ \text{N}}{(1.60 \times 10^{-19}\ \text{C})(5.00 \times 10^{7}\ \text{m/s})(2.00\ \text{T})} = 0.188$$

   and, $\theta = 10.8°$

4. $F = BIL = (1.2\ \text{T})(10\ \text{A})(2.0\ \text{m}) = 24\ \text{N}$

5. $F = BIL = (5.0 \times 10^{-5}\ \text{T})(10\ \text{A})(20\ \text{m}) = 1.0 \times 10^{-2}\ \text{N}$

   The force is westward (use right-hand rule)

6. $\tau_{max} = NBIA$, and $A = (4.00 \times 10^{-2}\ \text{m})(5.00 \times 10^{-2}\ \text{m}) = 2.00 \times 10^{-3}\ \text{m}^2$

   Thus, $I = \dfrac{\tau_{max}}{NBA} = \dfrac{(0.180\ \text{N m})}{(500)(0.650\ \text{T})(2.00 \times 10^{-3}\ \text{m}^2)} = 0.277\ \text{A}$

7. $r = \dfrac{mv}{qB} = \dfrac{(1.67 \times 10^{-27}\ \text{kg})(2.00 \times 10^{5}\ \text{m/s})}{(1.60 \times 10^{-19}\ \text{C})(0.200\ \text{T})} = 1.04 \times 10^{-2}\ \text{m} = 1.04\ \text{cm}$

8. $r = \dfrac{mv}{qB}$

   Thus, if B doubles, r will be reduced to $\frac{1}{2}$ its original value. So,

   $$r = \frac{1}{2}(1.77\ \text{m}) = 0.885\ \text{m}$$

9. We have $KE = \frac{1}{2}mv^2 = 1.60 \times 10^{-19}$ J

From which, $v = \sqrt{\frac{2(1.60 \times 10^{-19}\text{ J})}{9.11 \times 10^{-31}\text{ kg}}} = 5.93 \times 10^5$ m/s

and, $r = \frac{mv}{qB} = \frac{(9.11 \times 10^{-31}\text{ kg})(5.93 \times 10^5\text{ m/s})}{(1.60 \times 10^{-19}\text{ C})(0.200\text{ T})} = 1.69 \times 10^{-5}$ m

10. $m = \frac{qBr}{v} = \frac{(1.60 \times 10^{-19}\text{ C})(1.80\text{ T})(7.94 \times 10^{-3}\text{ m})}{(4.60 \times 10^5\text{ m/s})} = 4.97 \times 10^{-27}$ kg

11. (a) $B = \frac{\mu_o I}{2\pi r} = \frac{(4\pi \times 10^{-7}\text{ T m/A})(5\text{ A})}{2\pi(10^{-1}\text{ m})} = 10^{-5}$ T

(b) and (c) The method is the same as shown in (a). The answers are (b) 2×10^{-6} T, and (c) 5×10^{-7} T

12. From $\frac{F}{L} = \frac{\mu_o I_1 I_2}{2\pi d}$, we have

$$d = \frac{\mu_o I_1 I_2 L}{2\pi F} = \frac{(4\pi \times 10^{-7}\text{ T m/A})(10\text{ A})(10\text{ A})(0.50\text{ m})}{2\pi(1.0\text{ N})} = 10^{-5}\text{ m} = 10\ \mu\text{m}$$

It is highly unlikely that a wire of this radius could carry 10 A of current without melting.

13. Let us find the force on the wire carrying a current of 5.00 A. We shall designate this wire as wire 1. The magnitude of the magnetic field set up by wire 2 at the position of wire 1 is.

$$B_2 = \frac{\mu_o I_2}{2\pi r} = \frac{(4\pi \times 10^{-7}\text{ T m/A})(8.00\text{ A})}{2\pi(0.300\text{ m})} = 5.33 \times 10^{-6}\text{ T}$$

The force on wire 1 is

$$F_1 = B_2 I_1 L = (5.33 \times 10^{-6}\text{ T})(5.00\text{ A})(2.00\text{ m}) = 5.33 \times 10^{-5}\text{ N}$$ (directed toward the other wire)

14. The number of turns per unit length is

$$n = \frac{N}{L} = \frac{100}{0.150\text{ m}} = 6.67 \times 10^2\text{ m}^{-1}$$

and, $B = \mu_o nI$. From which,

$$I = \frac{B}{\mu_o n} = \frac{5.00 \times 10^{-5}\text{ T}}{(4\pi \times 10^{-7}\text{ T m/A})(667/\text{m})} = 5.97 \times 10^{-2}\text{ A} = 59.7\text{ mA}$$

15. We have $B = \mu_o nI$. Therefore, to increase the field B by a factor of 3, we must increase n by a factor of 3. But, $n = \frac{N}{L}$, or $N = nL$. Thus, if n increases by a factor of 3, N must increase by a factor of 3. If N(initial) = 100 then N(final) = 300 turns.

16. $B = \frac{mv}{rq} = \frac{(6.68 \times 10^{-27}\text{ kg})(10^4\text{ m/s})}{(1.60 \times 10^{-19}\text{ C})(3.00 \times 10^{-2}\text{ m})} = 1.39 \times 10^{-2}$ T (out of page)

20 INDUCED VOLTAGES AND INDUCTANCE

PROBLEMS

1. A magnetic field of strength 0.250 T is directed perpendicular to a circular loop of wire radius 30.0 cm. Find the magnetic flux through the area.

2. The magnetic flux through a loop consisting of two turns of wire is changing at the rate of 3 Wb/s. Find the magnitude of the induced emf in the loop.

3. A coil has a radius of 15.0 cm and is placed in an external magnetic field of strength 0.250 T with the plane of the coil perpendicular to the field direction. The field increases to 0.500 T in a time of 0.700 s. (a) Find the magnitude of the average induced emf during this time. (b) If the average current during this time was 0.800 A, find the resistance of the coil.

4. A magnetic field of 0.200 T exists within a solenoid of 500 turns and a diameter of 10.0 cm. How rapidly (that is, within what period of time) must the field be reduced to zero magnitude if the average magnitude of the induced emf within the coil during this time interval is to be 10.0 kV?

5. A circular wire loop of radius 0.50 m lies in a plane perpendicular to a uniform magnetic field of magnitude 0.40 T. If in 0.10 s the wire is reshaped from a circle into a square, but remains in the same plane, what is the magnitude of the average induced emf in the wire during this time?

6. A powerful electromagnet has a field of 1.6 T and a cross-sectional area of 0.20 m^2. If a coil of 200 turns with a total resistance of 20 Ω is placed around it, and then the power to the electromagnet is turned off in 0.020 s, what current is induced in the coil?

7. Coils rotating in a magnetic field are often used to measure unknown magnetic fields. As an example, consider a coil of radius 1.00 cm with 50.0 turns that is rotated about an axis perpendicular to the field at a rate of 20.0 Hz. If the maximum induced emf in the coil is 3.00 V, find the strength of the magnetic field.

8. A 500-turn circular coil of radius 20.0 cm is rotating about an axis perpendicular to a magnetic field of 0.100 T. What angular velocity will produce a maximum induced emf of 2.00 mV?

9. Calculate the magnetic flux through each turn of a 300-turn, 7.2-mH coil when the current in the coil is 10 mA.

10. Calculate the energy associated with the magnetic field of a 200-turn solenoid in which a current of 1.75 A produces a flux of 3.70×10^{-4} Wb in each turn.

11. Find the self-inductance of a coil if an emf of 100 mV is produced when the current changes from 0 to 1.5 A in a time of 0.30 s.

12. Consider a 100-turn rectangular coil of cross-sectional area 0.06 m^2 rotating with an angular velocity of 20 rad/s about an axis perpendicular to a magnetic field of 2.5 T. Plot the induced voltage as a function of time over one complete period of rotation.

13. A car with a 1.00-m-long radio antenna travels at 80.0 km/h in a place where the Earth's magnetic field is 5.00×10^{-5} T. What is the maximum possible induced emf in the antenna as it moves through the Earth's magnetic field?

14. An airplane with a wingspan of 40 m flies parallel to the Earth's surface, at a location where the downward component of the Earth's magnetic field is 0.60×10^{-4} T. At what speed would the plane have to fly to produce a difference in potential of 1.5 V between the tips of its wings? Is this likely to occur during a typical airplane flight?

CHAPTER 20 SOLUTIONS

1. $\Phi = BA = (0.250\text{ T})\pi(0.300\text{ m})^2 = 7.07 \times 10^{-2}\text{ T m}^2$

2. $\mathcal{E} = N\dfrac{\Delta\Phi}{\Delta t} = (2)(3\text{ T m}^2/\text{s}) = 6\text{ V}$

3. (a) We have, $\Delta\Phi = (\Delta B)A = (0.500\text{ T} - 0.250\text{ T})\pi(0.150\text{ m})^2 = 1.77 \times 10^{-2}\text{ T m}^2$

   Thus, the average value of the emf is

$$\mathcal{E} = \frac{\Delta\Phi}{\Delta t} = \frac{1.77 \times 10^{-2}\text{ T m}^2}{0.700\text{ s}} = 2.50 \times 10^{-2}\text{ V} = 25.2\text{ mV}$$

   (b) $R = \dfrac{\mathcal{E}}{i} = \dfrac{2.50 \times 10^{-2}\text{ V}}{0.800\text{ A}} = 3.16 \times 10^{-2}\ \Omega$

4. $\mathcal{E} = N\dfrac{\Delta\Phi}{\Delta t}$ and $\Delta\Phi = \Delta(BA) = (\Delta B)A$

   $(\Delta B) = (B_i - B_f) = 0.200\text{ T} - 0 = 0.200\text{ T}$

   we have $10.0 \times 10^3\text{ V} = (500)\dfrac{(0.200\text{ T})(\pi(0.050\text{ m})^2)}{\Delta t}$

   gives $\Delta t = 78.5\ \mu\text{s}$

5. The average induced emf is given by $\mathcal{E} = -N\left(\dfrac{\Delta\phi}{\Delta t}\right)$. Here $N = 1$, and

$$\Delta\Phi = B(A_{\text{square}} - A_{\text{circle}})$$

$$A_{\text{circle}} = \pi r^2 = \pi(0.50\text{ m}^2) = 0.7854\text{ m}^2$$

   Also, the circumference of the circle is $2\pi r = 2\pi(0.50\text{ m}) = 3.1416\text{ m}$. Therefore, each side of the square has a length

$$L = \frac{3.1416\text{ m}}{4} = 0.7854\text{ m} \quad \text{and} \quad A_{\text{square}} = L^2 = 0.6168\text{ m}^2$$

   Thus, $\Delta\Phi = (0.40\text{ T})(0.6168\text{ m}^2 - 0.7854\text{ m}^2)$, $\Delta\Phi = -0.067\text{ T m}^2$.

   The average induced emf is therefore:

$$\mathcal{E} - \frac{-0.067\text{ T m}^2}{0.10\text{ s}} = 0.67\text{ V}$$

6. $\Phi_i = B_i A = (1.6 \text{ T})(0.20 \text{ m}^2) = 0.32 \text{ T m}^2$ and $\Phi_f = 0$

Thus, $\mathcal{E} = N\dfrac{\Delta\Phi}{\Delta t} = (200)\dfrac{0.32 \text{ T m}^2 - 0}{0.02 \text{ s}} = 3200 \text{ V}$,

and $I = \dfrac{\mathcal{E}}{R} = \dfrac{3200 \text{ V}}{20\ \Omega} = 160 \text{ A}$.

7. We have, $\omega = 2\pi f = 2\pi(20.0 \text{ Hz}) = 126 \text{ s}^{-1}$

and $\mathcal{E}_{max} = NAB\omega$

Thus, $3.00 \text{ V} = (50.0)\pi(10^{-2} \text{ m})^2 B(126 \text{ s}^{-1})$

Which yields, $B = 1.52 \text{ T}$

8. $\mathcal{E}_{max} = NAB\omega$

$2.00 \times 10^{-3} \text{ V} = (500)\pi(0.200 \text{ m})^2(0.100 \text{ T})\omega$

and, $\omega = 3.18 \times 10^{-4} \text{ rad/s}$

9. $L = \dfrac{\Phi_{total}}{I} = \dfrac{N\Phi_{single\ loop}}{I}$

so $\Phi_{single\ loop} = \dfrac{LI}{N} = \dfrac{(7.2 \times 10^{-3} \text{ H})(10^{-2} \text{ A})}{300} = 2.4 \times 10^{-7} \text{ T m}^2$.

10. $L = \dfrac{N\Phi}{I} = \dfrac{(200)(3.70 \times 10^{-4} \text{ T m}^2)}{1.75 \text{ A}} = 4.23 \times 10^{-2} \text{ H}$

When $I = 1.75$ A in the coil, the energy stored in the magnetic field is

$$W = \frac{1}{2}LI^2 = \frac{1}{2}(4.23 \times 10^{-2} \text{ H})(1.75 \text{ A})^2 = 6.48 \times 10^{-2} \text{ J}$$

11. From $\mathcal{E} = L\dfrac{\Delta I}{\Delta t}$, we find

$$L = \frac{\mathcal{E}}{\frac{\Delta I}{\Delta t}} = \frac{100 \times 10^{-3} \text{ V}}{\frac{1.5 \text{ A}}{0.30 \text{ s}}} = \frac{0.10\text{V}}{5.0 \text{ A/s}} = 2 \times 10^{-2} \text{ H} = 20 \text{ mH}$$

12. The curve is a sine function with amplitude 300 V and period $\pi/10$ s.

13. The velocity is 80.0 km/h = 22.2 m/s. Then, if the antenna moves perpendicular to the Earth's field, the induced emf between the ends of the antenna is:

$$\mathcal{E} = -BLv = -(5.00 \times 10^{-5}\ \text{T})(1.00\ \text{m})(22.2\ \text{m/s}) = -1.11 \times 10^{-3}\ \text{V}$$

14. From, $\mathcal{E} = BLv$, we have: $v = \dfrac{\mathcal{E}}{BL} = \dfrac{1.5\ \text{V}}{(6.0 \times 10^{-5}\ \text{T})(40\ \text{m})}$, or v = 630 m/s = 1400 mi/h. This speed exceeds that of a typical airplane. Thus, the induced voltage will not reach 1.5 V.

21 ALTERNATING CURRENT CIRCUITS AND ELECTROMAGNETIC WAVES

PROBLEMS

*Indicates intermediate level problems.

1. For a particular ac generator, the output voltage is $v = 0.250\ V_m$ when $t = 0.002$ s. What is the operating frequency of the generator?

2. The current in a circuit containing a 4.00-μF capacitor is 0.300 A when connected to a generator whose rms output is 30.0 V. What is the frequency of the source?

3. A 100-μF capacitor is connected in series with an 18.0-Ω resistor and a 60.0-Hz ac voltage source. The voltage supplied to the circuit is 120 V. Determine the rms current in the circuit.

*4. A 20.0-Ω resistor, a 40.0-μF capacitor, and a 0.200-H inductor are connected in series with a 60.0-Hz, 70.0-V source. The inductor is found to have a resistance of 15.0 Ω. Find (a) the current in the circuit and (b) the voltage drop across the inductor.

5. A series ac circuit contains a 50.0-Ω resistor, a 15.0-μF capacitor, a 0.200-H inductor, and a 60.0-Hz generator that has an rms output of 90.0 V. Find the average power delivered to the circuit.

6. A series *RLC* circuit takes 108 W from a 110-V line. The applied voltage leads the voltage across the resistor by 37.0°. (a) What is the value of the current in the circuit? (b) What is the impedance of the circuit?

7. A transformer at a public utility reduces the voltage on a line from 360,000 V to 3600 V. The primary has 10,000 turns. (a) How many turns are there on the secondary? (b) If the current in the secondary is 600 A, find the current in the primary. Assume an ideal transformer.

8. A 250-mH inductor is connected across an ac source of $\Delta V_m = 90.0$ V. (a) At what frequency will the inductive reactance equal 20.0 Ω? (b) What is the rms value of current in the inductor at the frequency found in part (a)?

9. An electromagnetic wave in vacuum has an electric field amplitude of 150 V/m. Calculate the amplitude of the accompanying magnetic field.

*10. A radio transmitter broadcasts uniformly in all directions with an average power of 15.0 kW. Compute the maximum value of the electric field in its radio wave at the following distances from the transmitter: (a) 1.00 km, (b) 10.0 km, (c) 100 km.

11. The light from a 5.00-mW laser is spread out in a cylinder beam whose diameter is about 0.500 cm. What are the peak values of E and B in this beam?

12. The Moon is approximately 250,000 mi from us and the Sun is approximately 93,000,000 mi away. How much time is required for an electromagnetic wave to reach us from (a) the Sun and (b) the Moon?

CHAPTER 21 SOLUTIONS

1. We know that $\Delta v = \Delta V_m \sin 2\pi ft$. Thus,

$$\sin 2\pi ft = \frac{\Delta v}{\Delta V_m} = 0.250$$

From this, we find $2\pi ft = 0.253$ rad.

At $t = 0.002$ s,

$$f = \frac{0.253}{2\pi(0.002 \text{ s})} = 20.1 \text{ Hz}$$

2. $X_C = \frac{\Delta V}{I} = \frac{30.0 \text{ V}}{0.300 \text{ A}} = 100\ \Omega$

Therefore, $\frac{1}{2\pi fC} = 100\ \Omega$, or

$$f = \frac{1}{2\pi(100\ \Omega)(4.00 \times 10^{-6} \text{ F})} = 398 \text{ Hz}$$

3. We first calculate $X_C = \frac{1}{2\pi fC} = 26.53\ \Omega$. Then the impedance of the circuit is:

$$Z = \sqrt{R^2 + X^2{}_C} = \sqrt{(18.0)^2 + (26.53)^2} = 32.06\ \Omega$$

Therefore, $\Delta V = IZ$ yields: $I = \frac{\Delta V}{Z} = \frac{120 \text{ V}}{32.06\ \Omega} = 3.74$ A.

*4. (a) The total resistance in the circuit is the sum of the resistive element plus the resistance of the wire of the inductor. Thus,

$$R = 20.0\ \Omega + 15.0\ \Omega = 35.0\ \Omega$$

Also, $X_C = \frac{1}{2\pi fC} = \frac{1}{2\pi(60.0 \text{ Hz})(40.0 \times 10^{-6} \text{ F})} = 66.3\ \Omega$

and $X_L = 2\pi fL = 2\pi(60.0 \text{ Hz})(0.200 \text{ H}) = 75.4\ \Omega$.

Thus, $Z = \sqrt{(35.0\ \Omega)^2 + (75.4\ \Omega - 66.3\ \Omega)^2} = 36.2\ \Omega$

and $I = \frac{\Delta V}{Z} = \frac{70.0 \text{ V}}{36.2\ \Omega} = 1.94$ A

(b) $\Delta V_{\text{inductor}} = IZ_{\text{inductor}} = I\sqrt{R_L{}^2 + (X_L)^2} = (1.94 \text{ A})\ \sqrt{(15.0\ \Omega)^2 + (75.4\ \Omega)^2} = 149$ V

5. We find $X_L = 75.4\ \Omega$, $X_C = 177\ \Omega$, and $Z = 113\ \Omega$.

Thus, $I = \dfrac{\Delta V}{Z} = \dfrac{90.0\text{ V}}{113\ \Omega} = 0.795\text{ A}$

and $\tan\phi = \dfrac{X_L - X_C}{R} = \dfrac{75.4 - 177}{50.0} = -2.04$

and $\phi = -63.9°$

The power is found as $P = I\Delta V\cos\phi = (0.795\text{ A})(90.0\text{ V})\cos(-63.9°) = 31.6\text{ W}$

or $P = I^2R = (0.795)^2(50.0\ \Omega) = 31.6\text{ W}$

6. (a) $I = \dfrac{P}{\Delta V\cos\phi} = \dfrac{108\text{ W}}{(110\text{ V})\cos 37°} = 1.23\text{ A}$

(b) $Z = \dfrac{\Delta V}{I} = \dfrac{110\text{ V}}{1.23\text{ A}} = 89.5\ \Omega$

7. (a) From $\Delta V_s = \dfrac{N_s}{N_p}\Delta V_p$, we have

$$N_s = \frac{\Delta V_s}{\Delta V_p}N_p = \frac{3600}{360000}(10^4) = 100\text{ turns}$$

(b) From $P_{in} = P_{out}$, or $\Delta V_p I_p = \Delta V_s I_s$, we find

$$I_p = \frac{\Delta V_s}{\Delta V_p}I_s = \frac{3600}{360000}(600\text{ A}) = 6.00\text{ A}$$

8. (a) From $X_L = 2\pi fL$, we have

$$f = \frac{20.0\ \Omega}{2\pi(0.250\text{ H})} = 12.7\text{ Hz}$$

(b) $\Delta V = \dfrac{\Delta V_m}{\sqrt{2}} = \dfrac{90.0\text{ V}}{\sqrt{2}}$ and $I = \dfrac{\Delta V}{X_L} = \dfrac{90.0\text{ V}}{\sqrt{2}(20.0\ \Omega)} = 3.18\text{ A}$

9. $B = \dfrac{E}{c} = \dfrac{150\text{ V/m}}{3.00\times 10^8\text{ m/s}} = 5.00\times 10^{-7}\text{ T}$

10. $I = \dfrac{P}{A} = \dfrac{P}{4\pi r^2}$ and $I = \dfrac{E_{max}{}^2}{2\mu_o c}$

Thus, $E^2{}_{max} = \dfrac{\mu_o c}{2\pi r^2}$

$$P = \frac{(4\pi\times 10^{-7}\text{ Ns}^2/\text{C}^2)(3.00\times 10^8\text{ m/s})(15.0\times 10^3\text{ J/s})}{2\pi r^2} = \frac{(9.00\times 10^5\text{ N}^2\text{ m}^2/\text{C}^2)}{r^2}$$

10. (continued)
(a) at $r = 1.00$ km, the above gives $E_{max} = 0.949$ N/C

(b) at $r = 10.0$ km, $E_{max} = 0.0949$ N/C

(c) at $r = 100$ km, $E_{max} = 9.49 \times 10^{-3}$ N/C

11. $$I = \frac{P}{A} = \frac{5.00 \times 10^{-3}\ \text{W}}{\dfrac{\pi(5.00 \times 10^{-3}\ \text{m})^2}{4.00}} = 255\ \text{W/m}^2$$

and from $I = \dfrac{cB_{max}^{\ 2}}{2\mu_o}$, we have

$$B_{max} = \sqrt{\frac{2\mu_o}{c}(I)} = \sqrt{\frac{2(4\pi \times 10^{-7}\ \text{Ns}^2/\text{C}^2)(255\ \text{W/m}^2)}{(3.00 \times 10^8\ \text{m/s})}} = 1.46 \times 10^{-6}\ \text{T}$$

and, $E_{max} = cB_{max} = (3.00 \times 10^8\ \text{m/s})(1.46 \times 10^{-6}\ \text{T}) = 438$ N/C

12. $c = 3.00 \times 10^8$ m/s $= 1.86 \times 10^5$ mi/s

(a) $$t = \frac{d}{v} = \frac{93.0 \times 10^6\ \text{m i}}{1.86 \times 10^5\ \text{mi/s}} = 500\ \text{s} = 8.33\ \text{min}$$

(b) $$t = \frac{d}{v} = \frac{2.50 \times 10^5\ \text{m i}}{1.86 \times 10^5\ \text{mi/s}} = 1.34\ \text{s}$$

22 REFLECTION AND REFRACTION OF LIGHT

PROBLEMS

*Indicates intermediate level problems.

1. In Galileo's attempt to determine the speed of light, he and his companion were located on hilltops 9.0 km apart. What time interval between uncovering his lantern and seeing the light from his companion's lantern would Galileo have had to measure in order to obtain a value for the speed of light?

2. Find the speed of light in benzene.

3. The wavelength of sodium light in air is 589 nm. Find its wavelength in ethyl alcohol.

4. A light ray in air is incident on a water surface at an angle of 30.0° with respect to the normal surface. What is the angle of the refracted ray relative to the normal to the surface?

5. An underwater swimmer observes that a beam of sunlight in the water makes an angle of 31.0° with the vertical. Assuming the surface of the water is level and horizontal, determine the beam's angle as it enters the water.

6. A light ray initially in water enters a transparent substance at an angle of 37.0° with respect to the normal, and the transmitted ray is refracted at an angle of 25.0°. Calculate the speed of light in the transparent substance.

*7. You are standing with a mirror at the center of a giant clock. Someone at 12 o'clock shines a beam of light toward you, and you want to use the mirror to reflect the beam toward an observer at 5 o'clock. What should the angle of incidence be to achieve this?

*8. How far does a beam of light travel in water in the same time it takes it to travel 10.0 m in glass of index of refraction 1.50?

9. Calculate the critical angle for the following materials when surrounded by water: (a) zircon, (b) fluorite, (c) ice. Assume that $\lambda = 589$ nm.

10. A beam of light is incident at an angle of 37.0° on the surface of a block of transparent material. The angle of refraction is found to be 25.0°. What is the speed of light in the material?

11. The Sun is 10.0° above the horizon. If you are swimming beneath the surface of a pool of water, at what angle above the horizon does the Sun appear to be?

12. A material with index of refraction $n = 2.0$ is in the shape of a quarter circle of radius $R = 10$ cm and is surrounded by a vacuum (Fig. 22.1). A light ray, parallel to the base of the material, is incident from the left at a distance of $L = 5.0$ cm above the base and emerges out of the material at the angle θ. Determine the value of θ.

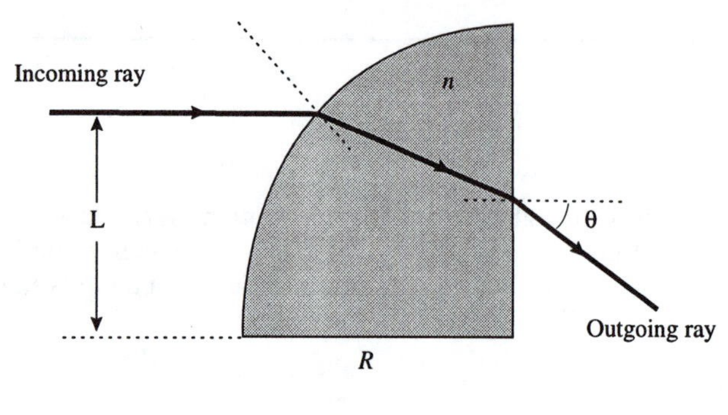

Figure 22.1

CHAPTER 22 SOLUTIONS

1. $t = \frac{d}{c} = \frac{1.8 \times 10^4 \text{ m}}{3.00 \times 10^8 \text{ m/s}} = 6.0 \times 10^{-5} \text{ s} = 60\ \mu\text{s}$

2. $v = \frac{c}{n} = \frac{3.00 \times 10^8 \text{ m/s}}{1.501} = 2.00 \times 10^8 \text{ m/s}$

3. $\lambda_{\text{medium}} = \frac{\lambda_{\text{vac}}}{n_{\text{medium}}} = \frac{589 \text{ nm}}{1.361} = 433 \text{ nm}$

4. We call medium 1 the air, and medium 2 is the water. Snell's law becomes

   $$n_2 \sin \theta_2 = n_1 \sin \theta_1$$

   $$\frac{4}{3} \sin \theta_2 = \sin 30.0°$$

   From which, $\sin \theta_2 = 0.375$, and $\theta = 22.0°$.

5. $\sin\theta_1 = \frac{n_2 \sin\theta_2}{n_1} = \frac{\frac{4}{3}\sin 31.0°}{1} = 0.687$

   $$\theta_1 = 43.4°$$

6. We find the index of refraction from Snell's law.

   $$n_2 \sin \theta_2 = n_1 \sin \theta_1$$

   $$n_2 \sin 25° = 1.33 \sin 37.0°$$

   $$n_2 = 1.90$$

   and $v = \frac{c}{n} = \frac{3.00 \times 10^8 \text{ m/s}}{1.90} = 1.58 \times 10^8 \text{ m/s}$

7. We have $\alpha = i + r = 2i$. (See sketch.)

   $$\alpha = \frac{5}{12} \text{ of a full circle} = \frac{5}{12}(360°) = 150°$$

   Thus, $i = \frac{150°}{2} = 75°$.

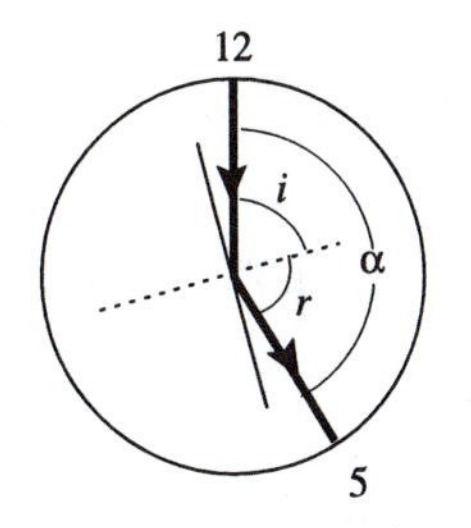

*8. In a time t, the light travels a distance d_w in water, where

$$d_w = v_w t = \frac{c}{n_w} t$$

In the same time, the distance traveled in glass is

$$d_g = v_g t = \frac{c}{n_g} t$$

Thus, $\frac{d_w}{d_g} = \frac{n_g}{n_w}$

Or, $d_w = \frac{1.50}{1.33}(10.0 \text{ m}) = 11.3 \text{ m}$

9. We use $\sin\theta_c = \frac{n_2}{n_1}$

(a) For zircon: $\sin \theta_c = \frac{1.333}{1.923} = 0.6934$, and $\theta_c = 43.9°$

(b) For fluorite: $\sin \theta_c = \frac{1.333}{1.434} = 0.9298$, and $\theta_c = 68.4°$

(c) For ice: $\sin \theta_c = \frac{1.333}{1.309} > 0$
Therefore, impossible. Total internal reflection cannot occur in this case.

10. By use of Snell's law, we find the index of refraction of the material.

$$n_2 \sin \theta_2 = n_1 \sin \theta_1$$

$$n_2 \sin 25° = (1.00) \sin 37.0°$$

and $n_2 = 1.424$

Thus, $v = \frac{c}{n} = \frac{3.00 \times 10^8 \text{ m/s}}{1.424} = 2.11 \times 10^8 \text{ m/s}$

11. The angle of incidence is 80.0°. We find the angle of refraction from Snell's law.

$$n_2 \sin \theta_2 = n_1 \sin \theta_1$$

$$1.33 \sin \theta_2 = (1.00) \sin 80°$$

gives, $\theta_2 = 47.6°$

Thus, with 47.6° as the angle between the refracted ray and the normal, we see that the angle between the ray and the horizontal is

$$90.0° - 47.6° = 42.4°$$

12. Observe that in the sketch the angle of incidence at point P is γ, and using triangle OPQ:

$$\sin \gamma = \frac{L}{R} = \frac{5.0 \text{ cm}}{10 \text{ cm}} = 0.50$$

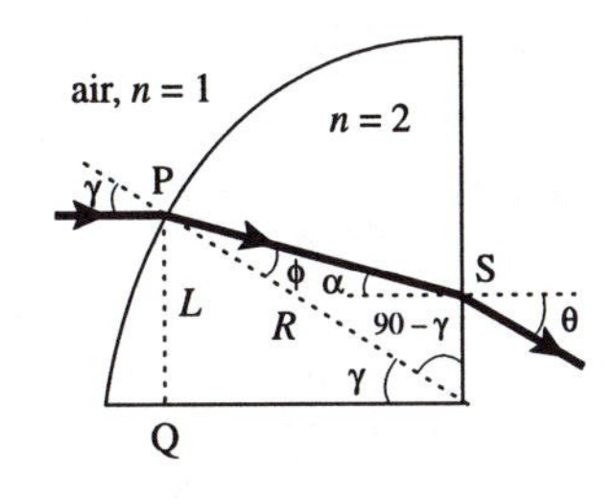

so $\gamma = 30°$.

Applying Snell's law at point P gives:

$2.0 \sin \phi = 1 \sin 30°$ or $\sin \phi = 0.25$ and $\phi = 14.5°$

Now use triangle OPS to see that:

$$\phi + (\alpha + 90°) + (90° - \gamma) = 180°$$

giving $\alpha = \gamma - \phi = 30° - 14.5° = 15.5°$ as the angle of incidence at point S. Applying Snell's law at point S gives:

$1 \sin \theta = 2.0 \sin(15.5°)$ and $\theta = 32°$

23 MIRRORS AND LENSES

PROBLEMS

1. Consider the case in which a light ray is incident on mirror 1 in Figure 23.1. The reflected ray is incident on mirror 2 and is subsequently reflected. Let the angle of incidence (with respect to the normal) on mirror 1 be 53°. Determine the angle between the ray incident on mirror 1 and the ray reflected on mirror 2.

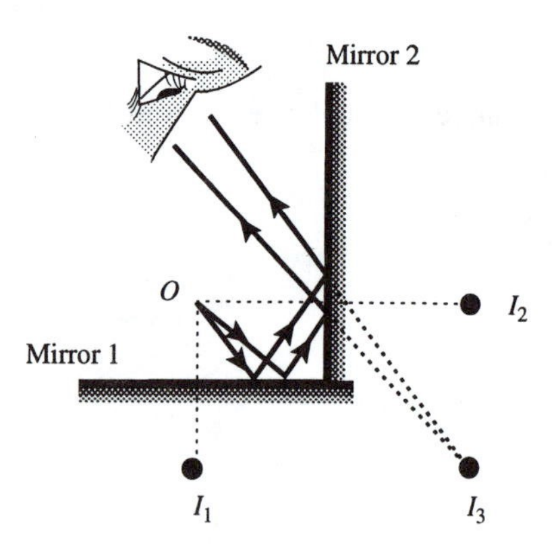

Figure 23.1

2. A goldfish is swimming in water inside a spherical plastic bowl of index of refraction 1.33. If the goldfish is 10.0 cm from the wall of the 15.0-cm-radius bowl, where does the goldfish appear to an observer outside the bowl?

3. A goldfish is swimming at 2.00 cm/s toward the right side of a rectangularly shaped aquarium tank. What is the apparent speed of the goldfish as measured by an observer looking in from outside the right side of the tank? The index of refraction for water is 1.33.

4. A virtual image is formed 20 cm from a concave mirror, having a radius of curvature of 40 cm. Find the position of the object.

5. A convex spherical mirror of focal length 15 cm is to form an image 10 cm from the mirror. Where should the object be placed?

6. When an object is moved along the principal axis of a thin lens, the height of the image is five times the height of the object if the object is at point A (to the left of the lens) and also if the object is at point B, 20.0 cm farther from the lens. (a) Is the lens converging or diverging? (b) What is the focal length of the lens?

7. A thin lens has a focal length of 25.0 cm. Locate the image when the object is placed (a) 26.0 cm and (b) 24.0 cm in front of the lens. Describe the image in each case.

8. A thin lens has a focal length of 225.0 cm. Locate the image when the object is placed as in Problem 7. Describe the image in each case. Determine the magnification.

9. An object is placed 30 cm from a lens and a virtual image is formed 15 cm from the lens. (a) What is the focal length of the lens? (b) Is the lens converging or diverging?

10. An object is placed 50.0 cm from a screen. Where should a converging lens with a 10.0-cm focal length be placed to form an image on the screen? Find the magnification(s).

11. A concave mirror has a focal length of 30.0 cm. (a) What is its radius of curvature? Locate and describe the image when the object is at (b) 100 cm and (c) 10.0 cm.

12. A planoconvex lens (n = 1.50) is flat on one side. What must be the radius of curvature of the curved side to produce a converging lens of focal length 20 cm?

13. Repeat Problem 12 for a planoconcave lens having a focal length of 220 cm.

CHAPTER 23 SOLUTIONS

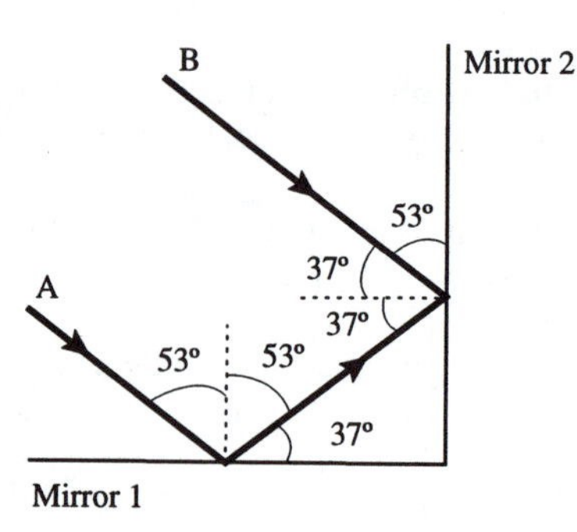

1. (See sketch.) Note from the sketch that both rays A and B make a 53° angle with the vertical. Thus, rays A and B are parallel to each other.

2. $\frac{n_1}{p} + \frac{n_2}{q} = \frac{n_2 - n_1}{R}$ becomes: $\frac{1.33}{10.0 \text{ cm}} + \frac{1.00}{q} = \frac{1.00 - 1.33}{-15.0 \text{ cm}}$

   from which $q = -9.00$ cm, or the fish appears to be 9.00 cm inside the wall of bowl.

3. For a plane surface,

   $$\frac{n_1}{p} + \frac{n_2}{q} = \frac{n_2 - n_1}{R} \text{ becomes: } q = -\frac{n_2 p}{n_1}$$

   Thus, if a change Δp occurs in the object's location, a change of magnitude $|\Delta q| = \frac{n_2 \Delta p}{n_1}$ will occur in the location of the image. If these changes occur in time Δt, the rates of change in position (i.e., the speeds) of the object and image are given by: $\frac{|\Delta q|}{\Delta t} = \left(\frac{n_2}{n_1}\right)\left(\frac{\Delta p}{\Delta t}\right)$. In our case, $\frac{\Delta p}{\Delta t} = 2.00$ cm/s, $n_1 = 1.33$ and $n_2 = 1.00$. Thus,

   $$v_{\text{image}} = \frac{\Delta q}{\Delta t} = \frac{1.00}{1.33}(2.00 \text{ cm/s}) = 1.50 \text{ cm/s}$$

4. We are given that $s' = -20$ cm, and for a concave mirror R and f are positive.

   Thus, $R = 40$ cm, and $f = \frac{R}{2} = 20$ cm

   $$\frac{1}{s} = \frac{1}{f} - \frac{1}{s'} = \frac{1}{20 \text{ cm}} - \frac{1}{-20 \text{ cm}} = \frac{1}{10 \text{ cm}}$$

   and $s = 10$ cm

   The object should be 10 cm in front of the mirror.

5. For a convex mirror f is negative.

   We also know that $s = -10$ cm because a convex mirror can only form virtual images of real objects. Thus,

   $$\frac{1}{x} = \frac{1}{f} - \frac{1}{s'} = \frac{1}{-15 \text{ cm}} - \frac{1}{-10 \text{ cm}} = \frac{1}{30 \text{ cm}}$$

   and $s = 30$ cm

   Thus, the object should be placed 30 cm in front of the mirror.

6. (a) Since the object is real, the lens must be converging, and the object must be inside the focal point in one case (virtual image) and outside in one case (real image). (When the object is outside the focal point of a diverging lens, the image is smaller than the object, not larger as specified.)

   (b) Using $M = \frac{-q}{p}$ we have: $q = -Mp$. The thin lens equation is then:

   $$\frac{1}{p} + \frac{1}{-Mp} = \frac{1}{f} \quad \text{or} \quad \frac{1}{p}\left(1 - \frac{1}{M}\right) = \frac{1}{f}$$

   When the object is inside the focal point, the image is erect, so $M = +5$, and $\frac{1}{p_1}\left(1 - \frac{1}{5}\right) = \frac{1}{f}$ gives: $p_1 = \frac{4}{5}f$.

   When the object is outside the focal point, the image is inverted, and $M = -5$. Thus,

   $$\frac{1}{(p_1 + 20.0\text{ cm})}\left(1 - \frac{1}{-5}\right) = \frac{1}{f}$$

   becomes $\left(\frac{1}{\frac{4}{5}f + 20.0\text{ cm}}\right)\frac{6}{5} = \frac{1}{f}$, from which $f = 50.0$ cm.

7. (a) $\frac{1}{s'} = \frac{1}{f} - \frac{1}{s} = \frac{1}{25.0\text{ cm}} - \frac{1}{26.0\text{ cm}}$ $\quad s' = 650$ cm

   The image is real, inverted, and enlarged.

   (b) $\frac{1}{s'} = \frac{1}{f} - \frac{1}{s} = \frac{1}{25.0\text{ cm}} - \frac{1}{24.0\text{ cm}}$ $\quad s' = -600$ cm

   The image is virtual, erect, and enlarged.

8. (a) $\frac{1}{s'} = \frac{1}{f} - \frac{1}{s} = \frac{1}{-25.0\text{ cm}} - \frac{1}{26.0\text{ cm}}$ $\quad s' = -12.8$ cm

   The image is erect, virtual, and diminished in size

   (b) $\frac{1}{s'} = \frac{1}{f} - \frac{1}{s} = \frac{1}{-25.0\text{ cm}} - \frac{1}{24.0\text{ cm}}$ $\quad s' = -12.2$ cm

   The image is erect, virtual, and diminished in size.

9. We are given that s = 30 cm and s = –15 cm. Thus, from $\frac{1}{s} + \frac{1}{s'} = \frac{1}{f}$.

$$\frac{1}{f} = \frac{1}{s} - \frac{1}{s'} = \frac{1}{30\text{ cm}} - \frac{1}{15\text{ cm}} = -\frac{1}{30\text{ cm}}$$

and f = –30 cm

The lens is diverging.

10. We know that: $p + q = 50.0$ cm, so $q = 50.0\text{ cm} - p$.

Also, $\frac{1}{p} + \frac{1}{q} = \frac{1}{f}$ becomes: $\frac{1}{p} + \frac{1}{(50.0\text{ cm} - p)} = \frac{1}{10.0\text{ cm}}$.

This reduces to: $p^2 - 50p + 500 = 0$, giving: $p = \frac{50.0\text{ cm} \pm 22.4\text{ cm}}{2}$.

Let us choose the plus sign in the above. We find, p = 36.2 cm, which yields an image distance of: q = 13.8, and magnification M = –0.382.

If we use the negative sign: p = 13.8 cm, q = 36.2 cm, and M = –2.62.

11. (a) $R = 2f = 60.0$ cm

(b) $\frac{1}{s'} = \frac{1}{f} - \frac{1}{s} = \frac{1}{30.0\text{ cm}} - \frac{1}{100\text{ cm}} = \frac{7}{300\text{ cm}}$ $\quad s' = 42.9$ cm

$$M = -\frac{s'}{s} = -\frac{42.9}{100} = -0.429$$

Thus, the image is real, inverted and located 42.9 cm in front of the mirror.

(c) $\frac{1}{s'} = \frac{1}{f} - \frac{1}{s} = \frac{1}{30.0\text{ cm}} - \frac{1}{10.0\text{ cm}} = -\frac{2}{30\text{ cm}}$ $\quad s' = -15.0$ cm

$$M = -\frac{s'}{s} = -\frac{-15.0}{10.0} = 1.50$$

The image is erect, virtual, and located 15.0 cm behind the mirror.

12. We use $\frac{1}{f} = (n - 1)\left(\frac{1}{R_1} - \frac{1}{R_2}\right)$

which becomes $\frac{1}{f} = (1.50 - 1)\left(\frac{1}{R_1} - 0\right) = \frac{1}{20\text{ cm}}$

from which, we find R_1 = 10 cm

13. $\frac{1}{f} = (n - 1)\left(\frac{1}{R_1} - \frac{1}{R_2}\right) = (1.50 - 1.0)\left(0 - \frac{1}{R_2}\right) = -\frac{1}{20\text{ cm}}$

from which, we find R_2 = 10 cm

24 WAVE OPTICS

PROBLEMS

1. Monochromatic light falls on a screen 1.50 m from two slits spaced 1.30 mm apart. The first- and second-order bright fringes are found to be 0.500 mm apart. What is the wavelength of the light?

2. A double slit with a spacing of 0.083 mm between the slits is 2.5 m from a screen. (a) If yellow light of wavelength 570 nm strikes the double slit, what is the separation between the zeroth- and first-order maxima on the screen? (b) If blue light of wavelength 410 nm strikes the double slit, what is the separation between the second- and fourth-order maxima? (c) Repeat parts (a) and (b) for the minima.

3. The yellow component of light from a helium discharge tube (λ = 587.5 nm) is allowed to fall on a plate containing parallel slits that are 0.200 mm apart. A screen is positioned so that the second bright fringe in the interference pattern is a distance equal to 10 slit spacings from the central maximum. What is the distance between the plate and the screen?

4. A thin film of glass (n = 1.52) with a thickness of 0.420 μm is viewed in white light at near-normal incidence. Visible light with what wavelength is most strongly reflected by the film when surrounded by air?

5. Waves from a radio station have a wavelength of 300 m. They arrive by two paths at a home receiver 20 km from the transmitter. One path is a direct path, and the second is by reflection from a mountain directly behind the home receiver. What is the minimum distance from the mountain to the receiver such that destructive interference occurs at the receiver? (Assume a phase change of 180° occurs on reflection from the mountain.)

6. An oil film 500 nm thick floats on water. It is illuminated with white light in the direction perpendicular to the film. What wavelengths will be strongly reflected in the range from 300 nm to 700 nm? Take n = 1.46 for oil.

7. One of the bright bands in Young's interference pattern is 12.0 mm from the central maximum. The screen is 119 cm from the pair of slits that serve as sources. The slits are 0.241 mm apart and are illuminated by the blue light from a hydrogen discharge tube (λ = 486 nm). How many bright lines are observed between the central maximum and the 12.0-mm position?

8. Infrared radiation of wavelength 12.2 mm passes through a single slit. (a) If the separation between the two first-order minima is 0.400 mm on a screen 84.0 cm away, what is the width of the slit? (b) If the width is reduced by half, what is the new separation between the first-order minima?

9. Light of wavelength 550 nm falls on a single slit of width 0.200 mm. Find the angle between the first dark bands on each side of the central maximum.

10. Determine Brewster's angle for light incident on benzene.

11. If the polarizing angle for cubic zirconia (ZrO_2) is 65.6°, what is the index of refraction for this material?

12. A soap bubble of index of refraction 1.40 strongly reflects both red and green colors in white light. What thickness of soap bubble allows this to happen? (In air, λ_{red} = 700 nm, λ_{green} = 500 nm).

13. Helium-neon laser light (λ = 632.8 nm) is sent through a 0.300-mm-wide single slit. What is the width of the central maximum on a screen 1.00 m in back of the slit?

14. Monochromatic light incident on a slit of width 0.0500 mm forms a diffraction pattern on a screen 2.00 m away. The second-order dark fringe is observed at an angle of 1.56° (see Figure 24.1). Calculate the wavelength of the light.

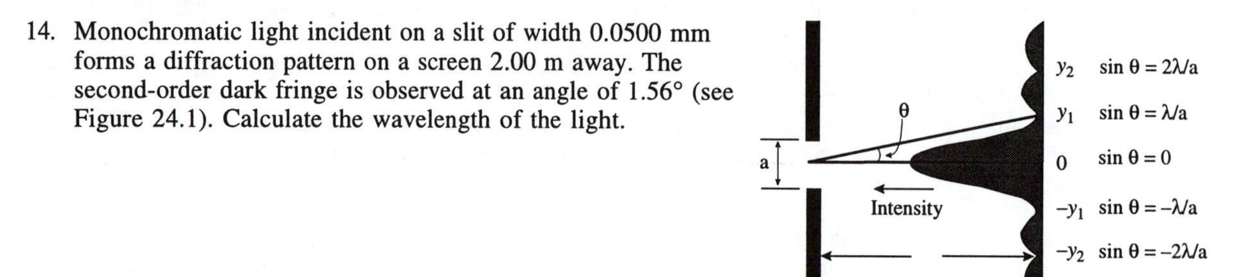

CHAPTER 24 SOLUTIONS

1. The position of the mth bright fringe is given by $y_m = \dfrac{m\lambda L}{d}$.

   Thus, $\Delta y = y_{m+1} - y_m = \dfrac{\lambda L}{d}$. From this, we find the wavelength as

   $$\lambda = \frac{(\Delta y)d}{L} = \frac{(5.00 \times 10^{-4}\text{ m})(1.30 \times 10^{-3}\text{ m})}{1.50\text{ m}} = 433\text{ nm}$$

2. (a) $\Delta y = [(y_{max})_{m=1} - (y_{max})_{m=0}] = \dfrac{\lambda L}{d}$, or

   $$\Delta y = \frac{(5.7 \times 10^{-7}\text{ m})(2.5\text{ m})}{8.3 \times 10^{-5}\text{ m}} = 1.7\text{ cm}$$

   (b) $\Delta y = [(y_{max})_{m=4} - (y_{max})_{m=2}] = 4\dfrac{\lambda L}{d} - 2\dfrac{\lambda L}{d}$, or

   $$\Delta y = 2\,\frac{(4.1 \times 10^{-7}\text{ m})(2.5\text{ m})}{8.3 \times 10^{-5}\text{ m}} = 2.5\text{ cm}$$

   (c) $\Delta y = [(y_{dark})_{m=1} - (y_{dark})_{m=0}] = \dfrac{\lambda L}{d}\left(1 + \dfrac{1}{2}\right) - \dfrac{\lambda L}{d}\left(0 + \dfrac{1}{2}\right) = 1.7$ cm
   (calculation same as in part (a).

   $\Delta y = [(y_{dark})_{m=4} - (y_{dark})_{m=2}] = \dfrac{\lambda L}{d}\left(4 + \dfrac{1}{2}\right) - \dfrac{\lambda L}{d}\left(2 + \dfrac{1}{2}\right) = 2.5$ cm
   (calculation same as in part (b).

3. We have $y_m = \dfrac{m\lambda L}{d}$, and we are given that $y_2 = 10d = 10(0.200\text{ mm}) = 2.00$ mm.

   Therefore, $2.00 \times 10^{-3}\text{ m} = \dfrac{2(587.5 \times 10^{-9}\text{ m})(L)}{2.00 \times 10^{-4}\text{ m}}$ yields:

   $L = 0.340\text{ m} = 34.0\text{ cm}$

4. There is a phase shift at the upper surface and none at the lower. We have

   $n_{film}(2t) + \dfrac{\lambda}{2} = m\lambda$ (for constructive interference)

   or $\left(m - \dfrac{1}{2}\right)\lambda = n_{film}(2t) = 2(1.52)(4.20 \times 10^{-7}\text{ m}) = 1277\text{ nm}$

   or $\lambda = \dfrac{1277\text{ nm}}{m - \frac{1}{2}}$ where $m = 1, 2, 3, \ldots$

   giving $\lambda = 2554$ nm, 851 nm, 511 nm, 365 nm (The only visible wavelength is 511 nm.)

5. For destructive interference, we have

$$\delta = 2n_f t + \frac{\lambda}{2} = \left(m + \frac{1}{2}\right)\lambda.$$

or, $2t = m\lambda$, and for the minimum distance, $m = 1$.

Thus, $t = \frac{\lambda}{2} = 150$ m

6. For constructive interference, we have

$$\delta = 2n_f t + \frac{\lambda}{2} = m\lambda.$$

and $\lambda = \frac{2n_f t}{\left(m - \frac{1}{2}\right)} = \frac{2(1.46)(500 \text{ nm})}{\left(m - \frac{1}{2}\right)}$

For $m = 1, 2, 3, 4$, etc. we have $\lambda = 2930$ nm, 973 nm, 584 nm, 417 nm, 324 nm, 265 nm etc.

But, between 300 nm and 700 nm, we find 324 nm, 417 nm, and 584 nm

7. The position of the mth order bright fringe on the screen is given by

$$y_{\text{bright}} = \frac{\lambda L}{d} m$$

Thus, if $y = 12$ mm, we have:

$$m = \frac{yd}{\lambda L} = \frac{(12.0 \times 10^{-3} \text{ m})(2.41 \times 10^{-4} \text{ m})}{(4.86 \times 10^{-7} \text{ m})(1.19 \text{ m})} = 5$$

Therefore, there must be four other bright fringes between this one and the central maximum.

8. The angles are small enough so that the approximation $\sin\theta = \tan\theta$ holds.

(a) Thus, $\sin\theta = \frac{m\lambda}{a}$, or $a = \frac{\lambda}{\sin\theta} = \frac{12.2 \times 10^{-6} \text{ m}}{2.38 \times 10^{-4}} = 5.12 \times 10^{-2} \text{ m} = 5.12$ cm

(b) if $a = \frac{1}{2}$ (old a) = 2.56 cm

then $\sin\theta = \frac{12.2 \times 10^{-6} \text{ m}}{2.56 \times 10^{-2}} = 4.76 \times 10^{-4}$,

and $\tan\theta = \frac{d}{L}$, so $d = (840 \text{ mm})(4.76 \times 10^{-4}) = 0.400$ mm

so the spacing between the first order minima = $2d = 0.800$ mm

9. At the first dark band, we have

$$\sin\theta = \frac{\lambda}{a} = \frac{5.50 \times 10^{-7}\text{ m}}{2.00 \times 10^{-4}\text{ m}} = 2.75 \times 10^{-3}\text{ m}$$

which yields $\theta = 0.158°$

This is the angle between the central maximum and the first dark band. Thus, the angle between the dark bands on each side of the central max is just twice this.

$$\alpha = 2\theta = 0.315°$$

10. We have $\tan\theta_p = \frac{n_2}{n_1} = \frac{1.501}{1} = 1.501$, and $\theta_p = 56.3°$

11. $n = \tan\theta_p = \tan(65.6°) = 2.2$

12. There is a phase shift at the first surface and none at the second.
Thus, $2n_f t + \frac{\lambda}{2} = m\lambda.$ (1)

For the same thickness to strongly reflect red (λ = 700 nm) and green (λ = 500 nm), we must have:

$$\left(m_R - \frac{1}{2}\right)(700\text{ nm}) = \left(m_g - \frac{1}{2}\right)(500\text{ nm})$$

which reduces to: $1.4m_R - m_g = 0.2$. The smallest integer values which satisfy this are: $m_R = 3$ and $m_g = 4$. Thus, equation (1) gives:

$$t = \frac{\left(3 - \frac{1}{2}\right)(700\text{ nm})}{2(1.40)} = 625\text{ nm}$$

13. $\sin\theta = \frac{\lambda}{a} = \frac{632.8 \times 10^{-9}\text{ m}}{3.00 \times 10^{-4}\text{ m}} = 2.109 \times 10^{-3}$

and $\tan\theta = \frac{d}{L}$ gives

$$d = L\tan\theta = L\sin\theta = (1.00\text{ m})(2.109 \times 10^{-3}) = 2.11\text{ mm}$$

Thus, the width of the central maximum = (distance between the first order minimum) = $2d$ = 4.22 mm

14. The position of the mth dark fringe is given by $\sin\theta = m\frac{\lambda}{a}$

So, $\lambda = \frac{a\sin\theta}{m}$. Thus, for the second order and at an angle of 1.56°, we have

$$\lambda = \frac{(5.00 \times 10^{-5}\text{ m})\sin 1.56°}{2.00} = 681\text{ nm}$$

25 OPTICAL INSTRUMENTS

PROBLEMS

*Indicates intermediate level problems.

1. The f-numbers of a simple camera are $f/4$, $f/8$, and $f/16$. What are the possible aperture openings for this camera if the focal length is 12 cm?

2. A certain lens forms a real image 15.0 cm to its right. The object is 20.0 cm to the left of the lens. If the lens has a diameter of 2.00 cm, find its f-number.

3. A camera is found to give proper film exposure when it is set at $f/16$ and the shutter is open for 1/32 s. Determine the correct exposure time if a setting of $f/8$ is used. Assume the lighting conditions are the same.

4. A certain farsighted person uses lenses with a power of 2.90 diopters. What is her near point?

5. A particular nearsighted person is unable to see objects clearly when they are beyond 100 cm (the far point of the eye). What power lens should be used to correct the eye?

6. A person required a lens with a power of –3 diopters in order to see distant objects clearly. What is the far point of this person's eye?

7. A lens with focal length 5.00 cm is used as a magnifying glass. (a) To obtain maximum magnification, where should the object be placed? (b) What is the magnification?

8. (a) What is the angular magnification of a lens having a focal length of 25 cm? (b) What is the magnification of this lens when the eye is relaxed?

9. A microscope is to have a magnification of 800. If the tube length is 15.0 cm and the focal length of the objective is 0.500 cm, what focal length eyepiece should be selected?

10. What is the magnification of a telescope that uses a 2.00-diopter objective and a 30.0-diopter eyepiece?

11. What is the minimum distance between two points that will permit them to be resolved a 1.00 km (a) using a terrestrial telescope with a 6.50-cm diameter objective (assume $\lambda = 550$ nm) and (b) using the unaided eye (assume a pupil diameter of 2.50 mm)?

12. A Michelson interferometer is used to measure an unknown wavelength. The mirror in one arm of the instrument is moved 0.120 mm as 481 dark fringes are counted. Determine the wavelength of the light used.

13. The second-order image formed by a diffraction grating having 6000 lines/cm is observed at an angle of 29.0°. What is the wavelength of the light used?

14. In a spectrum analyzed by a diffraction grating the fifth-order green maximum ($\lambda = 500$ nm) falls atop the fourth-order yellow maximum. What is the wavelength of the yellow light?

15. A diffraction grating is 3.00 cm wide and contains lines uniformly spaced at 775 nm. Determine the minimum wavelength difference that can be resolved in first order at 600 nm by this grating. Assume that the full grating width is illuminated.

16. The focal length of the lens in a simple camera is 10.0 cm, and the image formed on the film is to be 35.0 mm high. How far from the camera does a 2.00-m-tall person have to stand so that the person's image fits on the film?

17. In 1675 the Dutch biologist Anton van Leeuwenhoek, using a single lens probably of focal length 1.25 mm, discovered bacteria. What was the magnifying power of this early simple microscope?

18. Two radio telescopes are used in a special technique called long-baseline interferometry, which causes the two to act as a single telescope with an effective diameter of 1000 mi. At what distance from this telescope would two objects 250 000 mi apart be at the limit of resolution? Assume that the wavelength of the radio waves from these objects is about 1.00 m.

CHAPTER 25 SOLUTIONS

1. We have $D = \frac{f}{f\text{-number}} = \frac{12 \text{ cm}}{f}$

   f-number = 4 yields $\qquad D = \frac{12 \text{ cm}}{4} = 3 \text{ cm}$

   f-number = 8 yields $\qquad D = \frac{12 \text{ cm}}{8} = 1.5 \text{ cm}$

   f-number = 16 yields $\qquad D = \frac{12 \text{ cm}}{16} = 0.75 \text{ cm}$

2. The focal length is found from the thin lens equation as

   $$\frac{1}{f} = \frac{1}{s} + \frac{1}{s'} = \frac{1}{20.0 \text{ cm}} + \frac{1}{15.0 \text{ cm}}$$

   and from this $f = 8.57$ cm

   $$f\text{-number} = \frac{f}{D} = \frac{8.57 \text{ cm}}{2.00 \text{ cm}} = 4.29$$

3. The $f/8$ setting uses an aperture opening of twice the diameter of an $f/16$ setting. We have

   $$D_8 = \frac{\text{f}}{8} \quad \text{and} \quad D_{16} = \frac{\text{f}}{16}$$

   Thus, $\frac{D_8}{\text{D}_{16}} = \frac{16}{8} = 2$

   Therefore, since the amount of light entering the lens is proportional to the area of the opening, four times as much light will enter the camera.

   Thus, you should need $\frac{1}{4}$ as much exposure time, or

   $$t = \frac{1}{4}\left(\frac{1}{32}\text{ s}\right) = \frac{1}{128}\text{ s}$$

4. $f = \frac{1}{P} = \frac{1}{2.90} = 0.345 \text{ m} = 34.5 \text{ cm}$

   When the object is at $s = 25.0$ cm in front of the lens, the lens should form an erect, virtual image at the near point of the eye. From the thin lens equation,

   $$\frac{1}{s'} = \frac{1}{f} - \frac{1}{s} = \frac{1}{0.345 \text{ m}} - \frac{1}{0.250 \text{ m}}$$ which yields $s' = -0.909 \text{ m} = -90.9 \text{ cm}$

   Thus, the near point of the eye is at 90.9 cm from the eye.

5. The lens should form an erect, virtual image of the most distant objects (s = infinity) at the far point ($s' = -100$ cm). Thus,

$$\frac{1}{s} + \frac{1}{s'} = \frac{1}{f}$$

$$0 + \frac{1}{-100 \text{ cm}} = \frac{1}{f} \quad \text{and} \quad f = -100 \text{ cm} = -1.00 \text{ m}$$

and the power is $P = \frac{1}{f} = \frac{1}{-1.00} = -1.00$ diopter

6. $f = \frac{1}{P} = \frac{1}{-3 \text{ m}} = -0.333 \text{ m} = -33.3 \text{ cm}$

The location of the image formed of a very distant object is

$$\frac{1}{s} + \frac{1}{s'} = \frac{1}{f}$$

$$0 + \frac{1}{s'} = \frac{1}{-33.3 \text{ cm}} \quad \text{and} \quad s' = -33.3 \text{ cm}$$

The far point of the eye is at 33.3 cm.

7. The maximum magnification is obtainedwhen the virtual image is at the near point of the eye (25 cm in front of the eye).

$$\frac{1}{s} + \frac{1}{-25.0} = \frac{1}{5.00 \text{ cm}}$$

$$s = 4.17 \text{ cm}$$

and $M = 1 + \frac{25 \text{ cm}}{f} = 1 + \frac{25 \text{ cm}}{5.00 \text{ cm}} = 6.00$

8. (a) $M = 1 + \frac{25 \text{ cm}}{f} = 1 + \frac{25 \text{ cm}}{25 \text{ cm}} = 2$

(b) When the eye is relaxed $M = \frac{25 \text{ cm}}{f} = \frac{25 \text{ cm}}{25 \text{ cm}} = 1$

9. $M = -\frac{L}{f_o}\left(\frac{25 \text{ cm}}{f_e}\right) = -\frac{15.0 \text{ cm}}{0.500 \text{ cm}}\left(\frac{25.0 \text{ cm}}{f_e}\right) = -800$

and $f_e = 0.938$ cm

10. $f = \frac{1}{P}$, thus $f_o = \frac{1}{2.00} = 0.500 \text{ m} = 50.0 \text{ cm}$

and $f_e = \frac{1}{30.0} = 3.33 \times 10^{-2} \text{ m} = 3.33 \text{ cm}$

Therefore, $M = \frac{f_o}{f_e} = \frac{50.0 \text{ cm}}{3.33 \text{ cm}} = 15.0$

11. (a) $\theta_m = 1.22\frac{\lambda}{D} = 1.22\frac{5.50 \times 10^{-5} \text{ cm}}{6.50 \text{ cm}} = 1.03 \times 10^{-5} \text{ rad}$

Thus, $s = r\theta = (10^3 \text{ m})(1.03 \times 10^{-5} \text{ rad}) = 1.03 \times 10^{-2} \text{ m} = 1.03 \text{ cm}$

(b) $\theta_m = 1.22\frac{\lambda}{D} = 1.22\frac{5.50 \times 10^{-5} \text{ cm}}{0.250 \text{ cm}} = 2.68 \times 10^{-4} \text{ rad}$

Thus, $s = r\theta = (10^3 \text{ m})(2.68 \times 10^{-4} \text{ rad}) = 2.68 \times 10^{-1} \text{ m} = 26.8 \text{ cm}$

12. A fringe passes each time the mirror is moved a distance of half a wavelength. Thus, when 481 fringes pass, the mirror is moved a distance of

$$d = 481\,\frac{\lambda}{2} = 1.20 \times 10^{-4} \text{ m}$$

yielding $\lambda = 499$ nm

13. $d = \frac{1 \text{ cm}}{6000 \text{ lines}} = 1.667 \times 10^{-4} \text{ cm}$

and $n\lambda = d \sin\theta$ yields

$$\lambda = \frac{d \sin\theta}{n} = \frac{(1.667 \times 10^{-4} \text{ cm}) \sin 29.0°}{2} = 404 \text{ nm}$$

14. $n\lambda = d \sin\theta$ becomes

$$4\lambda_1 = d \sin\theta_4$$

and $5\lambda_2 = d \sin\theta_5$

but $\theta_5 = \theta_4$ (the fifth order line of the 500 nm wavelength coincides with the fourth order line of λ_1.

Therefore, $4\lambda_1 = 5\lambda_2$ and

$$\lambda_1 = \frac{5}{4}\lambda_2 = \frac{5}{4}\,500 \text{ nm} = 625 \text{ nm}$$

15. $N = \frac{\text{width}}{d} = \frac{3.00 \times 10^{-2}\text{ m}}{775 \times 10^{-9}\text{ m}} = 3.87 \times 10^{4}$ slits

Therefore, $R = \frac{\lambda}{\Delta\lambda} = mN$ or

$$\Delta\lambda = \frac{\lambda}{mN} = \frac{600\text{ nm}}{(1)(3.87 \times 10^{4})} = 0.0200\text{ nm}$$

16. From $h' = \frac{hq}{p}$, we have: $p = \left(\frac{h}{h'}\right)f = \frac{2.00\text{ m}}{3.50 \times 10^{-2}\text{ m}}(0.100\text{ m}) = 5.71\text{ m}.$

17. $M = \frac{25\text{ cm}}{f} = \frac{25\text{ cm}}{0.125\text{ cm}} = 200$ (assuming the eye is relaxed)

18. The limiting angle is:

$$\theta_{\text{m}} = 1.22\frac{\lambda}{D} = 1.22\frac{1.00\text{ m}}{1000\text{ mi}(1609\text{ m/mi})} = 7.582 \times 10^{-7}\text{ rad}$$

The distance at which two objects separated by $S = 250{,}000$ mi will be at this angular separation is:

$$r = \frac{S}{\theta_{\text{m}}} = \frac{(2.50 \times 10^{5}\text{ mi})}{7.582 \times 10^{-7}\text{ rad}} = 3.30 \times 10^{11}\text{ mi} = 5.31 \times 10^{11}\text{ km}$$

26 RELATIVITY

PROBLEMS

1. An unidentified flying object (UFO) flashes across the night sky. A UFO enthusiast on the top of Pike's Peak determines that the length of the UFO is 100 m along the direction of motion. If the UFO is moving with a speed of $0.900c$, how long is the UFO according to its pilot?

2. The nearest star to Earth is approximately 4.0 lightyears away. If you travel at 2.5×10^8 m/s in a spaceship, how long does it take to get there (a) according to an Earthbound observer and (b) according to an observer on the spaceship?

3. The nearest stars to Earth are approximately 4.00 lightyears away. How fast would you have to travel in a spaceship to cause that distance to shrink to 1.50 lightyears?

4. An atomic clock is placed on a jet airplane. The clock measures a time interval of 3600 s when the jet moves at 400 m/s. What corresponding time interval does an identical clock held by an observer on the ground measure? (*Hint*: For $v/c << 1$, use the approximation $\gamma \approx 1 + v^2/2c^2$.)

5. A 2.0-m long bobsled is traveling at 65 mi/h. What is the decrease in its length as seen by a stationary observer? (See hint in Problem 4.)

6. An astronaut in a spaceship travels at a speed of 0.8c away from an observer at rest on the Earth. He turns around and shines a beam of light toward the stationary observer. What does the stationary observer measure for the speed of light?

7. Find the rest energy of a proton in eV.

8. Find the speed of a particle whose total energy is 50% greater than its rest energy.

9. Assuming the mass of an electron can be converted completely into energy, how many electrons would have to be destroyed in order to run a 100-W lightbulb for one hour?

10. When 1.0 g of hydrogen combines with 8.0 g of oxygen, 9.0 g of water are formed. During this chemical reaction 2.86×10^5 J of energy are released. How much mass do the constituents of this reaction lose? Is the loss of mass likely to be detectable?

11. What are the momentum and kinetic energy of a proton moving with a speed of $0.980c$?

12. Energy reaches the upper atmosphere of the Earth from the sun at a rate of 1.79×10^{17} W. If all of this energy were absorbed by the Earth and converted into mass, how much would the mass of the Earth increase in one year?

13. Electrons are accelerated to an energy of 2.0×10^{10} eV in the 3.0-km-long Stanford Linear Accelerator. (a) What is the γ factor for the electrons? (b) What is the speed of the 20-GeV electrons? (c) How long does the accelerator appear to a 20-GeV electron?

CHAPTER 26 SOLUTIONS

1. $$\gamma = \frac{1}{\sqrt{1-\frac{v^2}{c^2}}} = \frac{1}{\sqrt{1-(0.900)^2}} = 2.294$$

The length measured by an observer in motion relative to the ship = 100 m = L_p. The length measured by pilot (at rest relative to ship) = $L = \gamma L_p$ = 2.294(100 m), or L_p = 229.4 m.

2. (a) The distance to the star as measured by an observer at rest on earth is 4 ly = 3.78×10^{16} m. The time to travel this distance according to an earthbound observer is:

$$T_p = \frac{d}{v} = \frac{3.78 \times 10^{16}\ \text{m}}{2.5 \times 10^8\ \text{m/s}} = 1.51 \times 10^8\ \text{s} = 4.8\ \text{years}$$

(b) This time, measured by the observer in the ship is given by $T = \gamma T_p$ where T_p is the time measured by the ship clock, and T is the time measured by an observer on earth (in motion relative to the ship clock). Thus,

$$T_p = \frac{T}{\gamma} = T\sqrt{1-\frac{v^2}{c^2}} = (4.8\ \text{y})\sqrt{1-\frac{(2.5 \times 10^8\ \text{m/s})^2}{(3 \times 10^8\ \text{m/s})^2}} = 2.7\ \text{years}$$

3. We are given L_p = 4.00 ly. Therefore, if L = 1.50 ly, then

$$L = L_p\sqrt{1-\frac{v^2}{c^2}}$$

becomes $\frac{L}{L_p} = \frac{1.50}{4.00} = 0.375 = \sqrt{1-\frac{v^2}{c^2}}$

From which, $\frac{v}{c} = 0.927$, or $v = 0.927\ c$

4. The time as measured by an observer in the plane at rest with respect to the clock is: T_p = 3600 s. For $v << c$, $\gamma \approx 1 + \frac{1}{2}\frac{v^2}{c^2}$. Therefore,

$$T = \gamma T_p \text{ becomes: } T \approx \left(1 + \frac{1}{2}\frac{v^2}{c^2}\right)T_p = T_p + \frac{1}{2}\frac{v^2}{c^2}T_p$$

or $T \approx 3600\ \text{s} + \frac{1}{2}\frac{(400)^2}{(3 \times 10^8)^2}(3600\ \text{s}) = 3600\ \text{s} + 3.2 \times 10^{-9}\ \text{s}$

5. $v = 65$ mi/h $= 29.1$ m/s. We are given $L_p = 2.0$ m.

Thus, $L = L_p\sqrt{1-\frac{v^2}{c^2}} = \frac{L_p}{\gamma}$, or $L_p = \gamma L = \gamma(L_p - \Delta L) = \gamma L_p - (\gamma\,\Delta L)$.

Therefore, $\Delta L = \frac{(\gamma - 1)}{\gamma} L_p$. For $v << c$, $\gamma \approx 1 + \frac{1}{2}\frac{v^2}{c^2}$, so

$$\Delta L \approx \frac{1 + \frac{1}{2}\frac{v^2}{c^2} - 1}{1 + \frac{1}{2}\frac{v^2}{c^2}} L_p = \frac{\frac{1}{2}\frac{v^2}{c^2}}{1 + \frac{1}{2}\frac{v^2}{c^2}} L_p = \frac{\frac{1}{2}\frac{(29.1)^2}{(3.00\times 10^8)^2}}{1 + \frac{1}{2}\frac{(29.1)^2}{(3.00\times 10^8)^2}} (2.00 \text{ m})$$

or $\Delta L \approx \frac{9.38\times 10^{-15}\text{ m}}{1 + 4.7\times 10^{-15}} = 9.4\times 10^{-15}$ m.

6. $$v_{LE} = \frac{v_{LS} + v_{SE}}{1 + \frac{v_{LS}v_{SE}}{c^2}} = \frac{-c + 0.8c}{1 + \frac{(-c)(0.8c)}{c^2}} = -c$$

7. $E_o = m_f c^2 = (1.67\times 10^{-27}\text{ kg})(3.00\times 10^8\text{ m/s})^2 = 1.50\times 10^{-10}\text{ J} = 9.39\times 10^8$ eV

8. We are given $E - E_o = 0.5E_o$

Thus, $E = 1.5E_o$

Also, $E = mc^2 = \gamma mc^2 = \gamma E_o$

So, $\gamma = 1.5 = \frac{1}{\sqrt{1-\frac{v^2}{c^2}}}$ and from this, $v = 0.745c$

9. The energy equivalent of one electron is

$$E_o = mc^2 = (9.11\times 10^{-31}\text{ kg})(3.00\times 10^8\text{ m/s})^2 = 8.20\times 10^{-14}\text{ J}$$

Thus $E = (\text{power})t = (100\text{ J/s})(3600\text{ s}) = 3.60\times 10^5$ J is the energy required to run the bulb for one hour.

$$\text{number electrons needed} = \frac{\text{total energy needed}}{\text{energy per electron}} = \frac{3.60\times 10^5\text{ J}}{8.20\times 10^{-14}\text{ J}} = 4.39\times 10^{18}$$

10. $E = 2.86\times 10^5$ J

Also, the mass-energy relation says that $E = mc^2$

Therefore, $m = \frac{E}{c^2} = \frac{2.86\times 10^5\text{ J}}{(3.00\times 10^8\text{ m/s})^2} = 3.18\times 10^{-12}$ kg

No, a mass loss of this magnitude (out of a total of 9 g) could not be detected.

11. At $0.95c$, $\gamma = 3.20$

$$p = \gamma mv = 3.20(1.67 \times 10^{-27}\ \text{kg})(0.950 \times 3.00 \times 10^{8}\ \text{m/s}) = 1.52 \times 10^{-18}\ \text{kg m/s}$$

$$KE = (\gamma - 1)mc^2 = (3.20 - 1)(1.67 \times 10^{-27}\ \text{kg})(3.00 \times 10^{8}\ \text{m/s})^2 = 3.31 \times 10^{-10}\ \text{J} = 2067\ \text{MeV}$$

12. The energy which arrives in one year is

$$E = (\text{power})t = (1.79 \times 10^{17}\ \text{J/s})(3.156 \times 10^{7}\ \text{s}) = 5.65 \times 10^{24}\ \text{J}$$

Thus, $m = \dfrac{E}{c^2} = \dfrac{5.65 \times 10^{24}\ \text{J}}{(3.00 \times 10^{8}\ \text{m/s})^2} = 6.28 \times 10^{7}\ \text{kg}$

13. (a) We are given that $E = 20$ GeV = 20,000 MeV. But, $E = \gamma E_o$, and $E_o = 0.511$ MeV for electrons. So,

$$\gamma = \frac{E}{E_o} = \frac{20000\ \text{MeV}}{0.511\ \text{MeV}} \quad \text{or} \quad \gamma = 3.9 \times 10^{4}$$

(b) $\gamma^2 = \dfrac{1}{1 - \dfrac{v^2}{c^2}}$ or $\dfrac{v^2}{c^2} = 1 - \dfrac{1}{\gamma^2} = 1 - \dfrac{1}{(39139)^2} = 1 - 6.53 \times 10^{-10}$

yielding $v = 0.9999999997c$.

(c) $L = \dfrac{L_p}{\gamma}$ yields $L = \dfrac{3000\ \text{m}}{39139} = 0.077\ \text{m} = 7.7\ \text{cm}$.

27 QUANTUM PHYSICS

PROBLEMS

1. Calculate the energy in electron volts of a photon in (a) the radio frequency range, 90.0 MHz, (b) the infrared range, 10^{13} Hz, and (c) the ultraviolet range, 10^{16} Hz.

2. A star that is moving away from the Earth at a speed of $0.280c$ is observed to emit radiation with a peak wavelength value of 500 nm. Determine the surface temperature of this star.

3. A metal has a work function of 2.00×10^{-19} J. If yellow light of wavelength 600 nm falls on the surface of the metal, find (a) the maximum kinetic energy of the ejected electrons and (b) the cutoff wavelength for the metal.

4. When light of wavelength 445 nm strikes a certain metal surface, the stopping potential is 70.0% of that which results when light of wavelength 410 nm strikes the same metal surface. Based on this information and the following table of work functions, identify the metal involved in the experiment.

Metal	Work Function (eV)
Cesium	1.90
Potassium	2.23
Silver	4.73
Tungsten	4.58

5. What is the shortest x-ray wavelength that can be produced with an accelerating voltage of 10.0 kV?

6. An electron initially at rest recoils after a head-on collision with a 0.14-keV photon. Determine the kinetic energy acquired by the electron.

7. X-rays of wavelength 0.200 nm are scattered from a block of carbon. If the scattered radiation is detected at 90.0° to the incident beam, find the Compton shift.

8. X-rays of wavelength 0.0710 nm undergo Compton scattering from free electrons in carbon. What is the wavelength of the photons, which are scattered at 90.0° relative to the incident direction?

9. Calculate the de Broglie wavelength of a 2000-kg car moving at 65.0 mi/h.

10. The position of an electron is known to a precision of 10^{-8} cm. What is the minimum uncertainty in the measurement of the electron's velocity?

11. The energy of an electron in a particular atom is approximately 2.00 eV. How long would it take to measure this energy to a precision of 1%?

12. The current in a photocell is cut off by a retarding potential of 0.92 V for radiation of wavelength 250 nm. Find the work function for the material.

13. The material in a photocell has a work function of 2.00 eV. When a retarding potential is applied, the cutoff wavelength is found to be 350 nm. What is the value of the retarding potential?

14. Find (a) the minimum energy of the photon required to produce a proton-antiproton pair and (b) the wavelength of this radiation.

CHAPTER 27 SOLUTIONS

1. (a) $E = hf = (6.63 \times 10^{-34}\text{ J s})(90.0 \times 10^{6}\text{ Hz}) = 5.97 \times 10^{-26}\text{ J} = 3.73 \times 10^{-7}\text{ eV}$

   (b) $E = hf = (6.63 \times 10^{-34}\text{ J s})(10^{13}\text{ Hz}) = 6.63 \times 10^{-21}\text{ J} = 4.14 \times 10^{-2}\text{ eV}$

   (c) $E = hf = (6.63 \times 10^{-34}\text{ J s})(10^{16}\text{ Hz}) = 6.63 \times 10^{-18}\text{ J} = 41.4\text{ eV}$

2. The radiation wavelength of $\lambda' = 500$ nm that is observed by observers on Earth is not the true wavelength, λ, emitted by the star because of the Doppler effect. In fact, the true wavelength is related to the observed wavelength using:

$$\frac{c}{\lambda'} = \frac{c}{\lambda}\sqrt{\frac{1-\frac{v}{c}}{1+\frac{v}{c}}} \quad \text{or} \quad \lambda = \lambda'\sqrt{\frac{1-\frac{v}{c}}{1+\frac{v}{c}}} = (500\text{ nm})\sqrt{\frac{1-0.28}{1+0.28}} = 375\text{ nm}$$

   The temperature of the star is obtained using $\lambda_{max}T = 0.2898 \times 10^{-2}$ m K,

   so $T = \dfrac{0.2898 \times 10^{-2}}{375 \times 10^{-9}} = 7.73 \times 10^{3}$ K.

3. (a) $KE_{max} = hf - \phi = \dfrac{(6.63 \times 10^{-34}\text{ J s})(3.00 \times 10^{8}\text{ m/s})}{6.00 \times 10^{-7}\text{ m}} - 2.00 \times 10^{-19}\text{ J}$

   $= 1.32 \times 10^{-19}\text{ J} = 0.822\text{ eV}$

   (b) $\lambda_c = \dfrac{hc}{\phi} = \dfrac{(6.63 \times 10^{-34}\text{ J s})(3.00 \times 10^{8}\text{ m/s})}{2.00 \times 10^{-19}\text{ J}} = 9.95 \times 10^{-7}\text{ m} = 995\text{ nm}$

4. From the photoelectric equation, we have:

$$e\Delta V_1 = E_{\gamma 1} - \phi \quad \text{and} \quad e\Delta V_2 = E_{\gamma 2} - \phi$$

   Since $\Delta V_2 = 0.700V_1$, then

$$e\Delta V_2 = 0.700(e\Delta V_1) = E_{\gamma 2} - \phi \quad \text{or} \quad (1 - 0.700)\phi = E_{\gamma 2} - E_{\gamma 1}$$

   and the work function is: $\phi = \dfrac{E_{\gamma 2} - 0.700E_{\gamma 1}}{0.3}$. The photon energies are:

$$E_{\gamma 1} = \frac{hc}{\lambda_1} = \frac{1243\text{ nm eV}}{410\text{ eV}} = 3.03\text{ eV} \quad \text{and} \quad E_{\gamma 2} = \frac{hc}{\lambda_2} = \frac{1243\text{ nm eV}}{445\text{ eV}} = 2.79\text{ eV}$$

   Thus, the work function is given by:

$$\phi = \frac{2.79\text{ eV} - 0.7(3.03\text{ eV})}{0.3} = 2.23\text{ eV}$$

   and we recognize this as characteristic of potassium.

5. $E_{\text{max}} = \dfrac{hc}{\lambda_{\text{min}}} = KE = qV$

so $\lambda_{\text{min}} = \dfrac{hc}{qV} = \dfrac{(6.63 \times 10^{-34} \text{ J s})(3.00 \times 10^{8} \text{ m/s})}{(1.60 \times 10^{-19} \text{ C})(10^{4} \text{ J/C})} = 1.24 \times 10^{-10} \text{ m} = 0.124 \text{ nm}$

6. The initial energy of the photon is: $E_\gamma = 6.2 \text{ keV} = \dfrac{hc}{\lambda}$, so $\lambda = 0.2$ nm.

Then, $\lambda' = \lambda + \dfrac{h}{mc}(1 - \cos\theta) = [(0.20 \text{ nm}) + (0.00243 \text{ nm})(1 - (-1))]$, or $\lambda' = 0.20486$ nm, giving

$E'_\gamma = = \dfrac{hc}{\lambda'} = 6.05 \text{ keV}$

Therefore, to conserve energy, the recoil kinetic energy of the electron must be:

$E_{\text{recoil}} = E_\gamma - E'_\gamma = 6.20 \text{ keV} - 6.05 \text{ keV} = 0.15 \text{ keV}.$

7. $\Delta\lambda = \dfrac{h}{mc}(1 - \cos\theta) = \dfrac{hc}{mc^2}(1 - \cos\theta) = \dfrac{hc}{E}(1 - \cos\theta)$

At $\theta = 90.0°$, $\Delta\lambda = \dfrac{1243 \text{ nm eV}}{0.511 \times 10^{6} \text{ eV}}(1 - 0) = 2.43 \times 10^{-3} \text{ nm}$

8. $\Delta\lambda = \dfrac{h}{mc}(1 - \cos\theta) = \lambda_c(1 - \cos\theta)$

When $\theta = 90.0°$, $\Delta\lambda = \lambda_c = 2.43 \times 10^{-3}$ nm

Thus, $\lambda = \lambda_0 + \Delta\lambda = 0.0710 \text{ nm} + 0.00243 \text{ nm} = 7.34 \times 10^{-2} \text{ nm}$

9. $\lambda = \dfrac{h}{p} = \dfrac{6.63 \times 10^{-34} \text{ J s}}{(2.00 \times 10^{3} \text{ kg})(65.0 \text{ mi/h})(0.447 \text{ m/s/mi h})} = 1.14 \times 10^{-38} \text{ m}$

10. $\Delta p \geq \dfrac{h}{4\pi(\Delta x)} = \dfrac{6.63 \times 10^{-34} \text{ J s}}{4\pi(10^{-10} \text{ m})} = 5.28 \times 10^{-25} \text{ kg m/s}$

so $\Delta v = \dfrac{\Delta p}{m} \geq \dfrac{5.28 \times 10^{-25} \text{ kg m/s}}{9.11 \times 10^{-31} \text{ kg}} = 5.79 \times 10^{5} \text{ m/s}$

11. $E = 2.00 \text{ eV} = 3.20 \times 10^{-19} \text{ J}$

Thus, one percent of $E = \dfrac{E}{100} = 3.20 \times 10^{-21} \text{ J}$

and from $(\Delta E)(\Delta t) \geq \dfrac{h}{4\pi}$

we have $\Delta t \geq \dfrac{h}{4\pi\Delta E} = \dfrac{6.63 \times 10^{-34} \text{ J s}}{4\pi(3.20 \times 10^{-21} \text{ J})} = 1.64 \times 10^{-14} \text{ s}$

12. $KE_{max} = e\Delta V = 0.92$ eV, and the photon energy is $E = hf = \dfrac{hc}{\lambda} = 4.97$ eV

    Thus from, $KE_{max} = E - \phi$, we find

    $$\phi = E - KE_{max} = 4.97 \text{ eV} - 0.92 \text{ eV} = 4.05 \text{ eV}$$

13. The retarding potential is just sufficient to stop the most energetic electrons liberated by the 350 nm light. Thus, $e\Delta V = KE_{max} = E - \phi$

    So, we use $\Delta V = \dfrac{E - \phi}{e}$ with $E = hf = \dfrac{hc}{\lambda} = 3.55$ eV, and $\phi = 2.00$ eV

    $$\Delta V = \frac{3.55 \text{ eV} - 2.00 \text{ eV}}{\text{e}} = 3.55 \text{ V} - 2.00 \text{ V} = 1.55 \text{ V}$$

14. (a) The minimum energy, E, of the photon required is twice the rest energy of a proton. Thus, $E = 2E_o = 2m_o c^2$.

    Therefore, $E = 2(1.67 \times 10^{-27} \text{ kg})(3.00 \times 10^8 \text{ m/s})^2 = 3.006 \times 10^{-10}$ J, or

    $$E = 1.88 \times 10^3 \text{ MeV}$$

    (b) $\lambda = \dfrac{hc}{E} = \dfrac{(6.63 \times 10^{-34} \text{ J s})(3.00 \times 10^8 \text{ m/s})}{3.006 \times 10^{-10} \text{ J}} = 6.62 \times 10^{-16}$ m

28 ATOMIC PHYSICS

PROBLEMS

1. Determine the wavelength of the following radiations of hydrogen: (a) the third-longest wavelength of the Paschen series; (b) the longest wavelength of the Balmer series; (c) the shortest wavelength (series limit) of the Paschen series; (d) the next-to-longest wavelength of the Lyman series. ($m = 1$ = Lyman series, $m = 2$ = Balmer series, $m = 3$ = Paschen series)

2. Show that the Balmer series formula (Equation 28.1) can be written as

$$\lambda = \frac{364.5n^2}{n^2 - 4} \text{ nm}$$

where $n = 3, 4, 5, \ldots$.

3. Show that $\dfrac{mk_e^2 e^4}{2\hbar^2} = 13.6$ eV

4. A photon is emitted from a hydrogen atom, which undergoes a transition from the $n = 3$ state to the $n = 2$ state. Calculate (a) the energy, (b) the wavelength, and (c) the frequency of the emitted photon.

5. Substitute numerical values into the equation below to determine a value for R_H.

$$R_H = \frac{m_e k_e^2 e^4}{4\pi c\hbar^3}$$

6. Calculate the radius of the first, second, and third Bohr orbits of hydrogen.

7. What wavelength radiation would ionize hydrogen when the atom is in its ground state?

8. Calculate the angular momentum of the electron in the $n = 3$ state of hydrogen.

9. Use the method illustrated in Example 28.7 to calculate the wavelength of the x-ray emitted from a molybdenum target ($Z = 42$) when an electron undergoes a transition from the L shell (n = 2) to the K shell ($n = 1$).

10. What is the radius of the first Bohr orbit in (a) He^+ and (b) Be^+?

11. List the possible sets of quantum numbers for electrons in the $3d$ subshell.

CHAPTER 28 SOLUTIONS

1. $\frac{1}{\lambda} = R\left(\frac{1}{m^2} - \frac{1}{n^2}\right)$

   $m = 1$ = Lyman series

   $m = 2$ = Balmer series

   $m = 3$ = Paschen series

   (a) When $n = 6$ $\quad \frac{1}{\lambda} = (1.0974 \times 10^7 \text{ m}^{-1})\left(\frac{1}{9} - \frac{1}{36}\right)$, and from this,

   $\lambda = 1.094 \times 10^{-6}$ m = 1094 nm

   (b) When $n = 3$ $\quad \frac{1}{\lambda} = (1.0974 \times 10^7 \text{ m}^{-1})\left(\frac{1}{4} - \frac{1}{9}\right)$, and from this,

   $\lambda =$ 656.1 nm

   (c) When $n = \infty$ $\quad \frac{1}{\lambda} = (1.0974 \times 10^7 \text{ m}^{-1})\left(\frac{1}{9} - \frac{1}{\infty}\right)$, and from this,

   $\lambda =$ 820.1 nm

   (d) When $n = 3$ $\quad \frac{1}{\lambda} = (1.0974 \times 10^7 \text{ m}^{-1})\left(\frac{1}{1} - \frac{1}{9}\right)$, and from this,

   $\lambda =$ 102.5 nm

2. Start with Balmer's equation: $\frac{1}{\lambda} = R_H\left(\frac{1}{2^2} - \frac{1}{n^2}\right)$, or $\lambda = \frac{(4n^2/R_H)}{(n^2 - 4)}$. Substituting $R_H = 1.0973732 \times 10^7 \text{ m}^{-1}$, we obtain

   $$\lambda = \frac{(3.645 \times 10^{-7} \text{ m})n^2}{n^2 - 4} = \frac{364.5n^2}{n^2 - 4} \text{ nm} \quad \text{where } n = 3, 4, 5, \ldots .$$

3. $$\frac{mk^2e^4}{2\hbar^2} = \frac{(9.11 \times 10^{-31} \text{ kg})(8.99 \times 10^9 \text{ N m}^2/\text{C}^2)^2(1.60 \times 10^{-19} \text{ C})^4}{2(1.055 \times 10^{-34} \text{ J s})^2} = 2.173 \times 10^{-18} \text{ J} = 13.6 \text{ eV}$$

4. (a) The energy of the photon is found as

$$E = E_i - E_f = \frac{-13.6\ \text{eV}}{n_i^2} - \frac{(-13.6\ \text{eV})}{n_f^2} = 13.6\ \text{eV}\left(\frac{1}{n_f^2} - \frac{1}{n_i^2}\right)$$

Thus, for $n = 3$ to $n = 2$ transition, $E = 13.6\ \text{eV}\left(\frac{1}{4} - \frac{1}{9}\right) = 1.89\ \text{eV}$

(b) $E = \frac{hc}{\lambda}$ and $\lambda = \frac{1243\ \text{nm eV}}{1.89\ \text{eV}} = 658\ \text{nm}$

(c) $f = \frac{c}{\lambda} = \frac{3.00 \times 10^8\ \text{m/s}}{6.58 \times 10^{-7}\ \text{m}} = 4.56 \times 10^{14}\ \text{Hz}$

5. $R = \frac{mk^2e^4}{4\pi c\hbar^3} = \frac{(9.11 \times 10^{-31}\ \text{kg})(8.99 \times 10^9\ \text{N m}^2/\text{C}^2)^2(1.60 \times 10^{-19}\ \text{C})^4}{4\pi(3.00 \times 10^8\ \text{m/s})(1.055 \times 10^{-34}\ \text{J s})^3} = 1.09243 \times 10^7\ \text{m}^{-1}$

6. $r_n = n^2 r_o = n^2(0.0529\ \text{nm})$

For $n = 1$, $r_1 = r_o = 0.0529$ nm

For $n = 2$, $r_2 = 4r_o = 4(0.0529\ \text{nm}) = 0.212$ nm

For $n = 3$, $r_3 = 9r_o = 9(0.0529\ \text{nm}) = 0.476$ nm

7. $E = \frac{13.6}{n^2}$ becomes for ($n = 1$) 13.6 eV,

and $\lambda = \frac{hc}{E} = \frac{1243\ \text{nm eV}}{13.6\ \text{eV}} = 91.4\ \text{nm}$

8. $L = n\hbar = 3(1.05 \times 10^{-34}\ \text{J s}) = 3.15 \times 10^{-34}\ \text{J s}$

9. $E_K = -(Z - 1)^2(13.6\ \text{eV}) = -(42 - 1)^2(13.6\ \text{eV}) = -(41)^2(13.6\ \text{eV}) = -22{,}862\ \text{eV}$

$$E_L = -(Z - 3)^2 \frac{(13.6\ \text{eV})}{2^2} = -(42 - 3)^2 \frac{(13.6\ \text{eV})}{4} = -(39)^2 \frac{(13.6\ \text{eV})}{4} = -5171\ \text{eV}$$

Thus, $\Delta E = E_L - E_K = -5171\ \text{eV} - (-22{,}862\ \text{eV}) = 17{,}691\ \text{eV} = E_\gamma$ (the energy of the x-ray emitted).

Therefore, $\lambda = \frac{hc}{E_\gamma} = \frac{1243\ \text{nm eV}}{17691\ \text{eV}} = 7.03 \times 10^{-2}\ \text{nm}$

10. $r_n = \frac{n^2 r_o}{Z}$ we have $r_n = \frac{0.529 \times 10^{-10}\ \text{m}}{Z}$

(a) for He^+ $r_n = \frac{0.529 \times 10^{-10}\ \text{m}}{2} = 0.265 \times 10^{-10}\ \text{m} = 0.265$ Å

(b) for Be^{3+} $r_n = \frac{0.529 \times 10^{-10}\ \text{m}}{4} = 0.132 \times 10^{-10}\ \text{m} = 0.132$ Å

11. In the 3d subshell, $n = 3$ and $l = 2$, We have

n	l	m_l	m_s
3	2	+2	+1/2
3	2	+2	−1/2
3	2	+1	+1/2
3	2	+1	−1/2
3	2	0	+1/2
3	2	0	−1/2
3	2	−1	+1/2
3	2	−1	−1/2
3	2	−2	+1/2
3	2	−2	−1/2

(A total of 10 states.)

29 NUCLEAR PHYSICS

PROBLEMS

*Indicates intermediate level problems.

1. Find the nucleus that has a radius approximately equal to one-half the radius of uranium, $^{238}_{92}U$.

2. Tritium has a half-life of 12.33 years. What percentage of the nuclei in a tritium sample will decay in five years?

3. Calculate the total binding energy for $^{40}_{20}Ca$.

4. Complete the following nuclear reactions:

   $^{27}_{13}Al + ^{4}_{2}He \rightarrow ? + ^{30}_{15}P$

   $^{1}_{0}n + ? \rightarrow ^{4}_{2}He + ^{7}_{3}Li$

5. Determine the difference between the binding energy of $^{3}_{1}H$ and $^{3}_{2}He$.

6. The half-life of a radioactive sample is 30.0 min. If you start with a sample containing 3.00×10^{16} nuclei, how many of these nuclei remain after 10.0 min?

7. How many radioactive atoms are present in a sample that has an activity of 0.20 μCi and a half-life of 8.1 days?

8. Calculate the energy released in the alpha decay of $^{232}_{92}U$.

9. Is it energetically possible for a $^{8}_{4}Be$ nucleus to decay spontaneously into two alpha particles? Explain.

10. Find the energy released in the fission reaction

    $$n + ^{235}_{92}U \rightarrow ^{98}_{40}Zr + ^{135}_{52}Te + 3n$$

    The atomic masses of the fission products are $^{98}_{40}Zr$, 97.9120 u, and $^{135}_{52}Te$, 134.9087 u.

11. Is it energetically possible for a $^{12}_{6}C$ nucleus to spontaneously decay into three alpha particles? Explain.

12. The radioactive isotope ^{198}Au has a half-life of 64.8 h. A sample containing this isotope has an initial activity of 40.0 μCi. Calculate the number of nuclei that will decay in the time interval between t_1 = 10 h and t_2 = 12 h.

13. In 2.40 s, a beam of x-rays produces 1.65×10^9 ion pairs in 2.20×10^{-3} kg of air. Determine the exposure and the exposure rate.

14. Calculate the radiation dose in rad supplied to 1.00 kg of water such that the energy deposited equals the water's thermal energy at 300 K. Assume that each molecule has a thermal energy kT.

CHAPTER 29 SOLUTIONS

1. From $r = r_o A^{1/3}$, the radius of uranium is $r_u = r_o(238)^{1/3}$

   Thus, if $r = \frac{1}{2} r_u$ then $r_o A^{1/3} = \frac{1}{2} r_o (238)^{1/3}$

   From which, $A = 30$

2. $\lambda = \frac{0.693}{T_{1/2}} = \frac{0.693}{12.33\ \text{y}} = 5.62 \times 10^{-2}\ \text{y}^{-1}$

   At $t = 5$ y; $\frac{N}{N_o} = e^{-\lambda t} = e^{-(5.62 \times 10^{-2}\ \text{y}^{-1})(5\ \text{y})} = 0.755$ (1)

   (1) is the fraction of the original which will remain after 5 y. Thus, the percent which has decayed = $(1 - 0.755) \times 100\% = 24.5\ \%$.

3. $\Delta m = 20m_H + 20m_n - m_{Ca} = 20(1.007825\ \mu) + 20(1.008665\ \mu) - 39.96259\ \mu = 0.367209\ \mu$.

   Therefore, $E_b = (\Delta m)c^2 = (0.367209\ \mu)(931.5\ \text{MeV}/\mu) = 342$ MeV

4. $^{27}_{13}\text{Al} + {}^{4}_{2}\text{He} \rightarrow {}^{1}_{0}\text{n} + {}^{30}_{15}\text{P}$

   $^{1}_{0}\text{n} + {}^{10}_{5}\text{B} \rightarrow {}^{4}_{2}\text{He} + {}^{7}_{3}\text{Li}$

5. For $^{3}_{1}\text{H}$ $\Delta m = 2m_n + m_H - m_{\text{tritium}} = (1.007825\ \mu) + 2(1.008665\ \mu) - 3.016049\ \mu = 0.009106\ \mu$

   and $E_b = (\Delta m)c^2 = 8.482$ MeV

   For $^{3}_{2}\text{He}$, $\Delta m = m_n + 2m_H - m_{\text{helium}} = 2(1.007825\ \mu) + (1.008665\ \mu) - 3.016029\ \mu = 0.008286\ \mu$

   and $E_b = (\Delta m)c^2 = 7.718$ MeV

   The difference in binding energy = 0.764 MeV, with the decreased binding energy in helium being mainly due to the increased Coulomb repulsion present.

6. The decay constant is $\lambda = \frac{0.693}{T_{1/2}} = \frac{0.693}{30.0\ \text{min}} = 2.31 \times 10^{-2}\ \text{min}^{-1}$.

   $N = N_o e^{-\lambda t} = (3.00 \times 10^{16}) e^{-(2.31 \times 10^{-2}\ \text{min}^{-1})(10.0\ \text{min})} = 3.00 \times 10^{16} e^{-0.231}$

   $= 3.00 \times 10^{16}(0.794) = 2.38 \times 10^{16}$ nuclei

7. $T_{1/2} = 8.1\ \text{days} = 7.0 \times 10^5$ s and $\lambda = \frac{0.693}{T_{1/2}} = \frac{0.693}{7 \times 10^5\ \text{s}} = 9.9 \times 10^{-7}\ \text{s}^{-1}$

   From $R = \lambda N$, $N = \frac{R}{\lambda}$

   If $R = 0.20\ \mu\text{Ci} = 7.4 \times 10^3$ decays/s then

   $N = \frac{7.4 \times 10^3\ \text{decays/s}}{9.9 \times 10^{-7}\ \text{s}^{-1}} = 7.5 \times 10^9$ nuclei

8. $^{232}_{92}U \rightarrow ^{4}_{2}He + ^{228}_{90}Th$

$E = (\Delta m)c^2 = [(M_{232U}) - (M_{4He} + M_{228Th})]$ (931.5 Mev/μ)

$=[232.037131\ \mu - (4.002602\ \mu + 228.028716\ \mu)]$ (931.5 Mev/μ)

$E = (0.005813\ \mu)$(931.5 Mev/μ) = 5.41 MeV

9. $^{8}_{4}Be \rightarrow ^{4}_{2}He + ^{4}_{2}He$

First, assume the decay will occur and compute the energy balance.

$E = (\Delta m)c^2 = [(M_{8Be}) - 2(M_{4He})]$ (931.5 Mev/μ)

$=[8.005305\ \mu - 2(4.002603)\ \mu]$ (931.5 Mev/μ)

$E = 92.2$ keV

Since $E > 0$ this means that the decay can occur spontaneously with an energy release.

10. $\Delta m = (m_n + M_U) - (M_{Zr} + M_{Te} + 3m_n)$, or

$\Delta m = (1.008665\ \mu + 235.043925\ \mu) - (97.9120\ \mu + 134.9087\ \mu + 3(1.008665\ \mu)$

$\Delta m = 0.20589\ \mu = 3.418 \times 10^{-28}$ kg, and $Q = \Delta mc^2 = 3.076 \times 10^{-11}$ J = 192 MeV

11. If the decay $^{12}_{6}C \rightarrow ^{4}_{2}He + ^{4}_{2}He + ^{4}_{2}He$ occurs, the Q value is

$Q = [M_C - 3(M_{He})]$ (931.5 Mev/μ) = $[12.00000\ \mu - 3(4.002603\ \mu)]$ (931.5 MeV/μ)

$= (-7.81 \times 10^{-3}\ \mu)$(931.5 MeV/$\mu$) = -7.27 MeV

Since the Q value is less than zero, the decay cannot occur spontaneously.

12. The decay constant of the isotope is

$$\lambda = \frac{0.693}{T_{1/2}} = \frac{0.693}{64.8\ \text{h}} = 1.069 \times 10^{-2}\ \text{h}^{-1} = 1.78 \times 10^{-4}\ \text{s}^{-1}$$

The initial number of nuclei is

$$N_o = \frac{R_o}{\lambda} = \frac{(40.0 \times 10^{-6})(3.7 \times 10^{10}\ \text{decays/s})}{1.78 \times 10^{-4}\ \text{s}^{-1}} = 8.30 \times 10^{9}\ \text{nuclei}$$

Now use $N = N_o e^{-\lambda t}$ to find the number present at $t = 10$ h.

$$N_{10} = (8.30 \times 10^9\ \text{nuclei})e^{-(1.069 \times 10^{-2}\ \text{h}^{-1})(10\ \text{h})} = 7.465 \times 10^9\ \text{nuclei}$$

12. (continued)
Similarly, the number present at 12 h can be found to be

$$N_{12} = 7.306 \times 10^9 \text{ nuclei}$$

Thus, the number which have decayed between $t = 10$ h and $t = 12$ h is

$$\Delta N = N_{10} - N_{12} = 1.60 \times 10^8 \text{ nuclei}$$

13. $$(\text{Volume of air present}) = \frac{\text{mass}}{\text{density}} = \frac{2.20 \times 10^{-3}\text{ kg}}{1.29\text{ kg/m}^3} = 1.705 \times 10^{-3}\text{ m}^3$$

or volume = 1705 cm^3

$$\text{exposure in Roentgens} = \frac{\dfrac{1.65 \times 10^9 \text{ ion pairs}}{1705 \text{ cm}^3}}{2.08 \times 10^9 \text{ ion pairs/cm}^3\text{/R}} = 4.65 \times 10^{-4}\text{ R}$$

$$\text{exposure rate} = \frac{4.65 \times 10^{-4}\text{ R}}{2.4\text{ s}} = 1.939 \times 10^{-4}\text{ R/s} = 0.698 \text{ Roentgens/h}$$

14. $$\text{The number of moles in one kg of water} = \frac{1000\text{ g}}{18\text{ g/mol}} = 55.6\text{ mol}$$

and the number of molecules in one kg

$$= (55.6\text{ mol})(6.02 \times 10^{23}\text{ molecules/mol}) = 3.34 \times 10^{25}\text{ molecules}$$

The thermal energy = (number of molecules)(kT)

$$= (3.34 \times 10^{25}\text{ molecules})(1.38 \times 10^{-23}\text{ J/K})(300\text{ K}) = 1.38 \times 10^5\text{ J}$$

The energy delivered by the radiation

$$= (\text{dose in rad})(10^{-2}\text{ J/kg rad})(1.00\text{ kg})$$

In order for this to equal the thermal energy, (dose in rad)(10^{-2} J/rad) = 1.38×10^5 J and

$$\text{dose in rad} = \frac{1.38 \times 10^5\text{ J}}{10^{-2}\text{ J/rad}} = 1.38 \times 10^7\text{ rad}$$

30 NUCLEAR ENERGY AND ELEMENTARY PARTICLES

PROBLEMS

1. A photon produces a proton-antiproton pair according to the reaction $\gamma \rightarrow p + p^-$. What is the frequency of the proton? What is its wavelength? (Assume that the proton and antiproton are left at rest after they are produced.)

2. In order to minimize neutron leakage from a reactor, the ratio of the surface area to the volume must be as small as possible. Assume that a sphere and a cube both have the same volume. Find the surface-to-volume ratio for (a) the sphere and (b) the cube. (c) Which of these shapes would have the minimum leakage?

3. A particle cannot generally be localized to distances smaller than its de Broglie wavelength. This means that a slow neutron appears to be larger than a fast neutron to a target particle because the slow neutron will probably be found over a larger volume of space. For a thermal neutron at room temperature (300 K) assume that its energy is given by $k_B T$ and find (a) its linear momentum and (b) the de Broglie wavelength. Compare this effective neutron size with both nuclear and atomic dimensions.

CHAPTER 30 SOLUTIONS

1. Assuming that the proton and the anitproton are left at rest after they are produced, the energy of the photon, E, must be

$$E = 2(938.3 \text{ MeV}) = 1876.6 \text{ MeV} = 3.00 \times 10^{-10} \text{ J}$$

Thus $E = hf = 3.00 \times 10^{-10}$ J and $f = \dfrac{3.00 \times 10^{-10} \text{ J}}{6.63 \times 10^{-34} \text{ J s}} - 4.53 \times 10^{23}$ Hz

and $\lambda = \dfrac{c}{f} = \dfrac{3.00 \times 10^{8} \text{ m/s}}{4.53 \times 10^{23} \text{ Hz}} - 6.62 \times 10^{-16}$ m

2. We are given: $V_{\text{sphere}} = V_{\text{cube}}$

If the sphere has a radius a and the cube has a length L on each side, we have: $\dfrac{4}{3}\pi a^3 = L^3$, or $L = (4\pi/3)^{1/3}(a)$.

(a) Sphere: $\dfrac{A}{V} = \dfrac{4\pi a^2}{\frac{4}{3}\pi a^3} = \dfrac{3}{a}$

(b) Cube: $\dfrac{A}{V} = \dfrac{6L^2}{L^3} = \dfrac{6}{\text{L}} = \dfrac{6}{(4\pi/3)^{1/3}(a)} = \dfrac{3.72}{a}$

(c) $\dfrac{A}{V}$ is lowest for the sphere. Thus, the sphere has a better shape to minimize leakage.

3. (a) $\dfrac{1}{2}mv^2 = kT$, or

$$v = \sqrt{\frac{2kT}{m}} = \sqrt{\frac{2(1.38 \times 10^{-23} \text{ J/K})(300 \text{ K})}{(1.675 \times 10^{-27} \text{ kg})}} = 2.22 \times 10^{3} \text{ m/s}$$

Thus, $p = mv = (1.675 \times 10^{-27} \text{ kg})(2.22 \times 10^{3} \text{ m/s}) = 3.72 \times 10^{-24}$ kg m/s.

(b) $\lambda = \dfrac{h}{p} = \dfrac{6.63 \times 10^{-34} \text{ J s}}{3.72 \times 10^{-24} \text{ kg m/s}} = 1.78 \times 10^{-10}$ m

The de Broglie wavelength is about the size of an atom (10^{-10} m), and about 10^5 times the size of a nucleus (10^{-15} m).